Wildflowers

of the Mountain West

Richard M. Anderson
JayDee Gunnell
Jerry L. Goodspeed

© 2012 by the University Press of Colorado

Published by Utah State University Press .
An imprint of University Press of Colorado
5589 Arapahoe Avenue, Suite 206C
Boulder, Colorado 80303

The University Press of Colorado is a proud member of

The Association of American University Presses.

AAUP 1937 2012

The University Press of Colorado is a cooperative publishing enterprise supported, in part, by Adams State College, Colorado State University, Fort Lewis College, Metropolitan State College of Denver, Regis University, University of Colorado, University of Northern Colorado, Utah State University, and Western State College of Colorado.

ISBN: 978-0-87421-895-4 (paper)
ISBN: 978-0-87421-896-1 (e-book)

Wildflowers should be enjoyed in their natural habitat. We discourage the removal or ingestion of any native plant. All edible and medicinal information found in this field guide is for historic reference only. Neither the authors nor the publisher is responsible for incorrect uses and/or claims of improper use.

Library of Congress Cataloging-in-Publication Data

Anderson, Richard M., 1966-
  Wildflowers of the Mountain West / Richard M. Anderson, JayDee Gunnell, Jerry L. Goodspeed.
    p. cm.
  Includes bibliographical references and index.
  ISBN 978-0-87421-895-4 (pbk.) — ISBN 978-0-87421-896-1 (e-book)
1. Wild flowers—West (U.S.)—Identification. 2. Mountain plants—West (U.S.)—Identification. 3. Wild flowers—West (U.S.)—Pictorial works. 4. Mountain plants—West (U.S.)—Pictorial works. I. Gunnell, JayDee, 1975- II. Goodspeed, Jerry L. III. Title.
  QK133.A53 2012
  582.130978—dc23
                    2012014197

Acknowledgments

We would foremost like to express gratitude to our families, who allowed us the freedom and time to go and discover. We love you and appreciate your patience with us.

We wish to acknowledge our employer, Utah State University Extension, and its administrators for generous funding support of this publication, and for allowing us to expand our own knowledge while serving and educating others regarding our natural resources. We literally have the best jobs in the world.

Many people have contributed to the success of this field guide. Thanks to the Intermountain Herbarium at Utah State University and the Ray J. Davis Herbarium at Idaho State University for allowing access to their extensive voucher collections and taxonomic expertise. Along these same lines we appreciate our graphic designers Jessica Buxton and Kyle Thornock, data input assistant Stacie Stone, digital photo editor Ryan Thompson, and our colleagues at Utah State University Botanical Center and Ogden Botanical Gardens. Our thanks also to Noelle E. Cockett, Vice President for Extension and Agriculture, Utah State University.

We would also like to acknowledge William A. Varga and Dr. Leila M. Shultz of Utah State University, and Dr. Steve Love of the University of Idaho for their continual encouragement during the writing of *Wildflowers of the Mountain West.*

Bannock County, Idaho, 2010, Elevation: 4,800 ft.

Contents

Largeleaf Balsamroot (*Balsamorhiza macrophylla*), Bannock County, Idaho, 2007, Elevation: 5,607 ft.

Introduction

In the summer of 2009, while leading a group of friends to enjoy mountain wildflowers, we noticed that many in our party had a desire to know the names and a little information about each of the flowers we were seeing. We also noticed that of the fifty pounds of books we lugged up the hill, most were too technical, had poor pictures, or were too cumbersome to make quick and correct identification of most wildflowers. This led to the three of us thinking about writing a field guide that would reduce the time spent identifying a flower and increase the time enjoying nature.

It is a normal instinct for us as humans to want to name nature's creations. This helps us to categorize and organize our world, which has been an endeavor for mankind since the early plant explorers, such as Linnaeus, Lewis and Clark, and others. The same feelings that drove those earlier naturalists drive us today.

In our effort to make identification easier for our friends, we also discovered that the whole nomenclature and taxonomy thing is harder to understand than a textbook written in some ancient language. The scientific name of some plants seems to change as often as the seasons, and what differentiates one species or variety from another can differ from one author to the next. We have strived (through voucher specimens, scientific literature, journals, and observations) to be consistent in scientific and common names with the majority of the botanists currently working with plants in the Mountain West region. However, as mentioned earlier, the naming of plants is a somewhat fluid science, and we are not always able to swim fast enough to stay out of the rapids.

We hope this field guide aids you in identifying wildflowers and allows you to spend less time in a book and more time experiencing the awesomeness and diversity of nature.

The Mountain West

The Mountain West, located between the Front Range of Colorado and the east slope of California's Sierra Nevada, is home to one of earth's most topographically elevated and biodiverse regions. According to numerous modern scholars, the plant species distributed throughout the Mountain West share a common genesis as a result of comparable climate and orogeny.[1] There is substantial evidence that indicates that climatic cycles during the geologic past have caused plants to migrate north to south, east to west, and along altitudinal gradients, as temperature and moisture have changed.

The orogenic events that elevated and shaped the Mountain West did not occur all at once; rather, a series of powerful uplifts and deformations occurred at different periods of geological time in response to the constant exertion of tectonic forces along the western edge of the North American plate. The result, after millions of years, was a system of mountains with extreme relief and general north-to-south parallelism–the Rocky Mountains (including the Southern Rockies, the Middle Rockies, and the Wyoming Basin), the Intermontane Plateau (including the Great Basin, the Uinta Basin, the Snake River Plain, and the Colorado Plateau), and the Pacific Mountains (Sierra Nevada).

Nowhere are these parallel ranges more evident than in the heart of the Intermontane Plateaus of Nevada and western Utah, where sagebrush basins give rise to alpine ranges in a rhythmic pattern across the region. These mountain systems collectively make up a portion of the North American *Cordillera*;[2] one of the largest mountain systems in the world.

The Southern Rockies are believed to have originated during the late Cretaceous and early Paleocene, some 66 million years ago.

---

[1] Refers to the forces and events that contribute to mountain-building processes over time.

[2] The North American Cordillera includes the vast mountainous trough between the Pacific Coast Ranges and the Front Range of Colorado and all of the mountain ranges from Central America to the Arctic Circle.

Rocky Mountain National Park, Grand County, Colorado, 2011, Elevation: 12,000 ft.

Wildflowers of the Mountain West

Malad Valley, Box Elder County, Utah, 2010, Elevation: 4,500 ft.

Today they make up a region that covers much of western Colorado, portions of southeastern Wyoming, and northern New Mexico. The Southern Rockies erupt from the prairies to the east and give way to the Colorado Plateau and Uinta Basin to the west. To the north they indiscernibly blend into the Middle Rockies, separated only by the Wyoming Basin. Of the western mountain systems, it is the highest on average, with fifty peaks exceeding 14,000 feet. At 14,433 feet, Mount Elbert, located in the Sawatch Range of central Colorado, is the highest point.

The Middle Rockies are centered in western Wyoming, with ranges extending northward into southwestern Montana and southward into northern Utah. Broad valleys, formed from rivers downcutting through Tertiary sediments (which covered the region millions of years ago), characterize and separate the individual ranges. The Yellowstone River, which flows northeastward from Yellowstone National Park through Montana, marks the boundary between the Middle and Northern Rockies. The flora, however, is not constrained by this arbitrary demarcation and individual species obliviously transition both to the north and south of this anthropocentric[3] boundary. The

---

3   Anthropocentrism is the human tendency to perceive ourselves as the central entity of the universe.

Green River, where it wraps around the Uinta Mountains to the east and south, marks the southern boundary. The Wasatch Range in northern Utah is significant because it denotes not only the western boundary of the Middle Rockies but also the eastern rim of the Great Basin.

The Intermontane Plateau stretches from western Colorado to the eastern foothills of the Sierra Nevada and is sandwiched between the Northern and Middle Rockies and the Sonoran and Mojave Deserts. The Colorado Plateau covers much of eastern Utah and western Colorado and extends southward into Arizona and New Mexico. The isolated La Sal Mountains in Utah and the San Juan Mountains in Colorado are important refugia[4] for plant species of Rocky Mountain origin that cannot survive on the Plateau floor.

Ninety percent of the water that reaches the Colorado Plateau is drained by the Colorado River, but the Great Basin is a vast, cold desert where no water escapes. It is bounded by the Sierra Nevada on the west and Middle Rockies on the east and resides wholly within what many scholars dub the *Intermountain West*. The region is characterized by some 200 individual block-fault mountain ranges separated by narrow valleys. In general, these ranges are lower and drier in relation to the Rocky Mountain system. A more natural definition of the Great Basin that takes its flora into account extends its northern boundary to include the Snake River Plain of southern Idaho and eastern Oregon. The Mojave and Sonoran Deserts mark the southern boundary. The White Mountains, situated on the California/Nevada border, and the Spring Mountains and Sheep Range near Las Vegas, Nevada, are additional isolated mountain ranges where numerous plant species are of Rocky Mountain origin.

The Sierra Nevada, in relation to the Southern Rockies, is much younger. It began to rise during the Eocene and Oligocene, some 30 million years ago. The greatest uplift occurred, however, during the middle Pliocene, about 3–4 million years ago. The Sierra Nevada, a narrow range between 40 and 80 miles in width, extends north to south along the eastern border of California for approximately 430 miles. The southern end is the Garlock Fault, where it intersects with the San Andreas Fault, near Tehachapi Pass. The northern end is Fredonyer Pass, west of Susanville. The east face of the Sierra Nevada consists of a steep block fault escarpment, which, in

---

4   Isolated assemblages of plant species that were once widespread.

a distance of approximately twelve miles, can change in elevation from 4,500 feet at the valley floor to heights of more than 12,000 feet. It is on this east face that approximately 40 percent of the plant species exhibit a common origin with those found as far east as the Southern Rockies. The highest peak in the Sierra Nevada is Mt. Whitney, at 14,505 feet, located on the border between Inyo and Tulare Counties.

The Climate

To cultivate a stronger understanding of why plants grow where they do, it helps to consider that the fundamental principle governing plant distribution is climate. The term *climate* denotes a sustained pattern of weather over a long period of time. These sustained climatic patterns, in turn, act according to broad (macro) and local (micro) geographic scales. Both of these scales are ultimately a function of temperature, which in turn is dependent upon the intensity and quality of solar energy (i.e., latitude,[5] elevation, exposure, and season).

The macroclimate of the Mountain West is predominantly continental;[6] however, environmental factors such as land form, large bodies of water, cloud cover, or wind, further modify the predominant weather patterns to create localized microclimates. For example, the maritime microclimate of the Sierra Nevada is a consequence of its proximity to the moderating effects of the Pacific Ocean. The Intermontane Plateau, on the other hand, lies in the rain shadow created by the Sierra Nevada. The result is a drier, semiarid microclimate. The role of elevation, as it relates to temperature regulation, is exemplified in the Rocky Mountains. Air masses moving over the Intermontane Plateau eventually collide with and are forcefully uplifted over the elevated peaks, resulting in adiabatic cooling[7] and release of moisture. Because of this, the Rocky Mountain system enjoys a temperate microclimate that is cooler and wetter in comparison to the Intermontane Plateau.

---

[5] According to Arthur Cronquist, one mile of latitude equates to four feet of altitude. Assuming that all environmental factors are the same, two points of equal elevation, separated by 75 miles of latitude, will have a climate difference equivalent to a 300-foot increase in elevation and a difference of 1°F.

[6] Continental climates are characterized by long, snowy winters, summer drought, short growing seasons, low ambient humidity, drastically fluctuating temperatures, intense solar radiation, shallow soils, and strong winds.

[7] Adiabatic cooling is a decrease in temperature at a rate of 1°F for each 300-foot increase in elevation.

Tundra Communities Trailhead, Rocky Mountain National Park, Grand County, Colorado, 2011, Elevation: 12,050 ft.

Temperature directly influences the rate of plant growth, including bloom time. While latitudinal warming marks the beginning of each growing season, elevational increases act to lengthen the overall period during which a particular species may bloom. For this reason a specific wildflower can be found blooming first at lower latitudes and elevations, and at progressively higher latitudes and elevations as latitudinal and elevational warming catches up.

The Flora

Besides climate, plant distribution can be further modified by the accumulating influences of orogeny, fauna, fire, and man. These governing forces are not mutually exclusive but rather interact in a concerted, compounding fashion to randomly (or, in consideration of human activity, not so randomly) catalyze speciation, migratory patterns, evolutionary adaptation, and, in some instances, extinction.

The earliest botanical perspectives considered plant distribution to be a static phenomenon. By the late eighteenth century, however, the advent of standardized botanical names and the acceleration of global exploratory activities precipitated a rapid accumulation of knowledge regarding global climate, soil, topography, and their effect on the distributional patterns of the world's plant species.

Kiger Gorge Overlook, Steens Mountain, Harney County, Oregon, 2011, Elevation: 8,910 ft.

By the time Alexander von Humboldt, the preeminent German naturalist/explorer, wrote his 1805 *Essay on the Geography of Plants*, it was broadly accepted that there was a significant correlation between climate and vegetation with regard to elevation and latitude. It was the French botanist Joseph Pittonde Tournefort who observed". . .that in ascending mountains we meet successively with vegetations that represent those of successively higher latitudes."[8] It was Humboldt who first stated that temperature decreases as elevation increases, which today we call adiabatic cooling.

In 1859, Charles Darwin published his monumental work *On the Origin of Species by Means of Natural Selection, or the Preservation of Favoured Races in the Struggle for Life*. In it he laid forth a sound argument addressing the questions regarding speciation, migration, and evolutionary adaptation. As a result of these works and those of many others, nineteenth-century botanists were able to conceptualize and articulate a new scientific idea—namely that distributional patterns of plants are in perpetual flux.

Owing to the enormous complexity of synthesizing a justifiable floristic boundary for such a vast region as the Mountain West, we have looked to the monumental 1985 paper by Daniel Axelrod and Peter Raven, "Origins of the Cordilleran Flora." According to these

---

[8]   Joseph D. Hooker, "On Geographic Distribution," *The Pharmaceutical Journal and Transactions*. 12 (1881-82): 511.

View of Grand Teton from Rendezvous Peak, Grand Teton National Park, Teton County, Wyoming, 2010, Elevation: 10,889 ft.

scholars, the Cordilleran flora (what we call the Mountain West flora) extends from the east slope of the Sierra Nevada to the eastern front of the Southern Rockies, includes the Snake River Plain to the north, and shifts quickly into the desert species of the Sonoran and Mojave to the south.

Although a summary of the complexities of geologic time in relation to the genesis of the Cordilleran flora is well beyond the scope of this field guide, it will help to know that lowland, frost-sensitive species from the flora south of the Mountain West, and the cold-hardy flora to the north, initially migrated into the Middle and Southern Rockies, colonized westward across the Intermontane Plateau as a consequence of decreased precipitation, and eventually reached the east slope of the Sierra Nevada. This happened because plant species were able to migrate into the Mountain West as favorable climatic and orogenic changes created new, suitable habitat. The reverse has also occurred. As the distributions of plant species expanded and contracted, only those individuals that possessed the genetic potential to adapt to the new environmental conditions remained. This process has repeated itself many times.

The Mountain West flora consists of individual plant species that have sorted, over time, into stable and distinctive multispecies assemblages along elevational gradients, called communities.

These gradients are traditionally referred to as foothill, montane, subalpine, and alpine, in ascending order. One will notice that species of the subalpine (8,000 to 11,000 feet) and alpine (> 11,000 feet) vegetation communities appear to be derivatives of or identical to numerous species in the modern northern flora, as in the genera *Pedicularis*, *Ranunculus*, *Saxifraga*, and *Silene*. On the other hand, plant species of the montane (6,000 to 9,000 feet) and foothills (4,000 to 6,000 feet) vegetation communities, as in the genera *Allium*, *Astragalus*, *Eriogonum*, and *Penstemon*, demonstrate a much stronger southern affinity.

The present distribution of plants in the Mountain West is an example of how climate interacts with numerous modifying factors to develop unique vegetation communities over time. This field guide introduces 130 of the most common wildflowers that can be easily seen from roadways, identified along nature trails, or experienced in the region's many national parks.

Mountain West Region

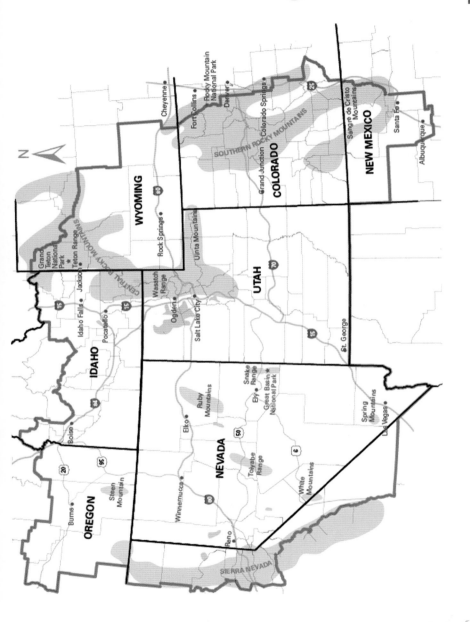

Basic Flower Shapes

Composite

Daisy-like

Dome-like

Simple

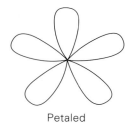

Petaled

Bulb-like

Tubular

Bell-shaped

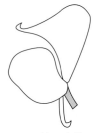

Hood-like

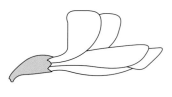

Pea-like

Flower Arrangement

Single

Solitary

Cluster

Umbel

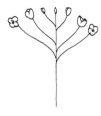

Corymb

Cyme

Along the Stalk

Spike

Raceme

Panicle

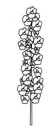

Bottlebrush-like
(tight spike or raceme)

Simple Flower Parts

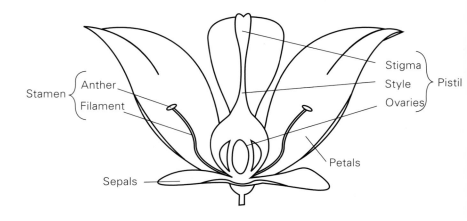

Stamen { Anther
         Filament

Stigma
Style
Ovaries
} Pistil

Petals

Sepals

Composite Flower Parts

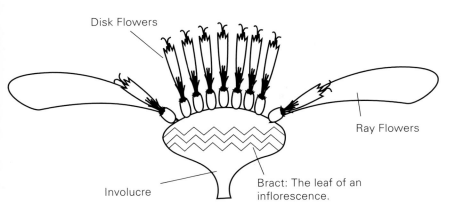

Disk Flowers

Ray Flowers

Involucre

Bract: The leaf of an inflorescence.

Simple Leaf Shapes

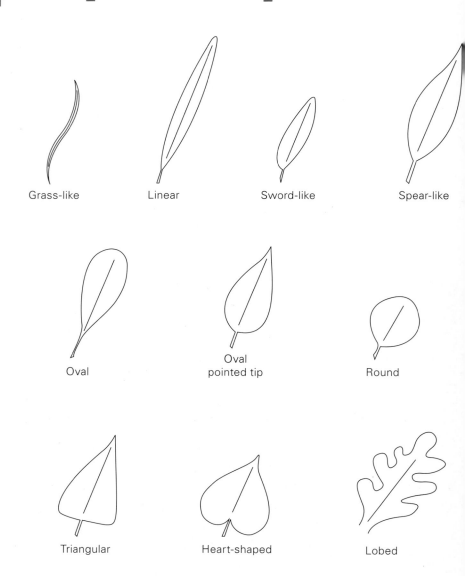

Grass-like

Linear

Sword-like

Spear-like

Oval

Oval
pointed tip

Round

Triangular

Heart-shaped

Lobed

Compound Leaf Shapes

Palm-like
(palmately compound)

Ladder-like
(pinnately compound)

Fern-like
(pinnate or bipinnate)

Leaf Arrangement

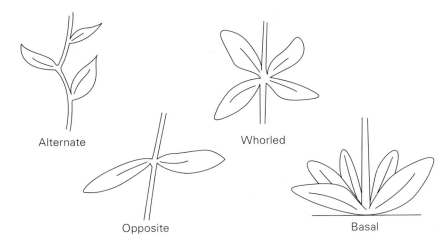

Alternate

Whorled

Opposite

Basal

Family Name
(Common - Scientific)

Common Name
Scientific name

Plant Size
General range of height

Flower
General description including shape, arrangement, color, and size

Bloom Time
Average bloom time (dependent on elevation and latitude)

Form/Foliage
General description including leaf shape, leaf arrangement, color, size, and other defining characteristics

Habitat
Environment where the wildflower is typically found

Elevation Range
An altitudinal range based on known field collections and observations

Look-Alike
Other plants that often resemble or are confused with this species

Map
Known distribution according to herbarium records and field observations (based on county lines)

Pea Family - Fabaceae

Northern Sweetvetch
Hedysarum boreale

Plant Size   6 to 24 inches

Flower   Clusters of pea-like flowers alternate along a single stalk (raceme) and resemble large folded butterfly wings

                  Color: Pink to magenta
                  Size: 1/2 to 3/4 inch long

Bloom Time   June to July

Form/Foliage   Mounding habit with leaves (green above and gray beneath) that are made up of 7 to 15 smaller leaflets (pinnately compound) that are arranged opposite each other along a main stem (petiole)

Habitat   Dry hillsides, sagebrush communities, and dry, open slopes

Elevation Range   4,100 to 9,100 feet

Look-Alike   Western Sweetvetch (*Hedysarum occidentale*)

Look-Alike drooping flowers

How to Use This Field Guide

Historically, Native Americans ate the roots, which supposedly taste like licorice.

Flora Fact
Informative and fun tidbits about this wildflower

Flower
Typical characteristics of individual flowers or overall arrangement (inflorescence)

Detail Insert
Close-up of leaf and/or flower

Plant Form or Habitat
The plant in a natural setting

WHITE FLOWERS

Spiny Phlox (*Phlox hoodii*), Bannock County, Idaho, 2010
Elevation: 4,669 ft.

Common Yarrow
Achillea millefolium

Plant Size	1 to 2 feet
Flower	Arranged in tight, flat-topped clusters (corymbs)
	Color: White, occasionally pink with white centers Size: 1/4 to 1/2 inch wide
Bloom Time	May to September
Form/Foliage	Upright habit with aromatic, fern-like, finely dissected, gray-green leaves that alternate along the stems
Habitat	Broad range including dry hillsides, sagebrush communities, and disturbed sites
Elevation Range	4,200 to 11,600 feet
Look-Alike	Douglas's Dustymaiden (*Chaenactis douglasii*) p. 6

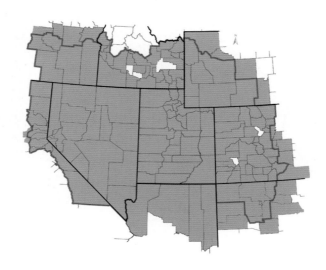

Historically, Native Americans used the rhizomes of this plant as a mild, local numbing agent for toothaches.

Pearly Everlasting
Anaphalis margaritacea

Plant Size	8 to 36 inches
Flower	Clusters of dome-shaped flowers resembling cotton swabs are actually daisy-like (composite) blossoms that lack outer petals (ray flowers). The white petal-like structures are modified leaves (bracts).
	Color: White with yellow centers
	Size: 1/4 inch wide (cluster)
Bloom Time	July to September
Form/Foliage	Upright habit with gray-green, fuzzy (pubescent) spear-like leaves that connect directly to the stem (sessile) and alternate along it
Habitat	Open hillsides, disturbed sites, and woodlands
Elevation Range	4,250 to 10,950 feet
Look-Alike	Pussytoes (*Antennaria* spp.) p.158

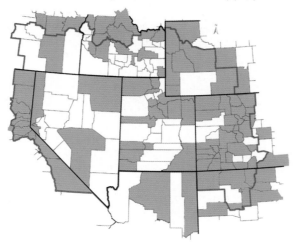

The dried flowers are often used in floral arrangements.

Douglas's Dustymaiden
Chaenactis douglasii

Plant Size	12 to 18 inches
Flower	Clusters of 40 to 50 tubular flowers (rayless composite) with protruding forked flower parts (styles) that resemble a pincushion
	Color: White to pale pink Size: About 1 inch wide (cluster)
Bloom Time	June to September
Form/Foliage	Upright and airy habit with gray-green, fern-like alternating leaves that create a distinct "frilly" appearance
Habitat	Dry, gravelly slopes
Elevation Range	4,000 to 9,500 feet
Look-Alike	Common Yarrow (*Achillea millefolium*) p. 2

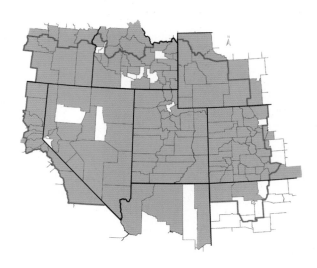

Also called Bride's Bouquet or False Yarrow.

Engelmann's Aster
Eucephalus (Aster) engelmannii

Plant Size   1 1/2 to 3 feet

Flower   Loose clusters of solitary daisy-like flowers
                          (composites) with 15 to 20 wide outer petals
                          (ray flowers)

                          Color: White to pink
                          Size: 1 1/2 to 2 1/2 inches wide

Bloom Time   July to September

Form/Foliage   Upright habit with narrow, smooth, spear-like
                          leaves that alternate along the stems

Habitat   Open woodlands, moist meadows, and stream
                          banks

Elevation Range   6,500 to 9,900 feet

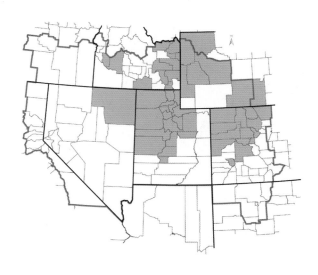

Asters have several overlapping rows of leaf-like structures (bracts) at the base of the flowers (involucre), much like shingles on a roof, whereas fleabane daisies have only one to two rows.

Douglas's Catchfly
Silene douglasii

Plant Size	8 to 24 inches
Flower	Loose clusters (cymes) of notched 5-petaled flowers emerging from distinct papery cylindrical structures (calyx tubes) Color: White Size: 3/4 inch wide
Bloom Time	June to July
Form/Foliage	Spindly habit with oval (pointed tip), gray-green, hairy (pubescent) leaves that are found opposite each other on the stems
Habitat	Dry hillsides and gravelly slopes
Elevation Range	5,000 to 11,000 feet
Look-Alike	There are over 70 different varieties of *Silene* in North America.

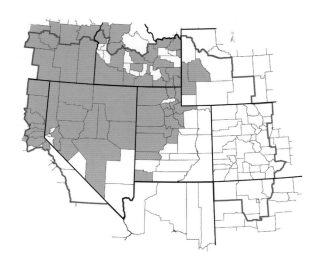

Insects may become "trapped" in the sticky resin produced by this plant, giving it the common name of Catchfly.

Richardson's Geranium
Geranium richardsonii

Plant Size   16 to 30 inches

Flower   Solitary 5-petaled, distinctly veined flowers
                        borne atop loosely branched stalks

                        Color: White with pink veins
                        Size: 1 to 1 1/2 inches wide

Bloom Time   May to September

Form/Foliage   Mounding habit with dark green, deeply
                        dissected, palm-like (palmately compound)
                        leaves that are primarily basal

Habitat   Moist meadows, open woodlands, and
                        sagebrush communities

Elevation Range   5,500 to 11,000 feet

Look-Alike   Sticky Geranium
                        (*Geranium viscosissimum*) p. 172

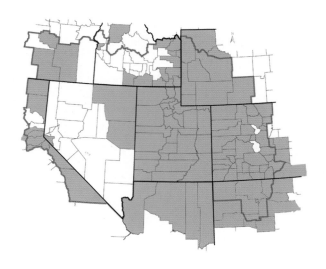

The beak-like seed (style) drills itself into the soil as it coils and uncoils.

Sego Lily
Calochortus nuttallii

Plant Size	8 to 16 inches
Flower	Large, bell-shaped flowers consisting of 3 rounded petals and a yellow throat, painted with a purple band on each individual petal
	Color: White with purple bands Size: 2 to 3 inches wide
Bloom Time	June to July
Form/Foliage	Leaves are slender, grass-like, and primarily basal
Habitat	Dry hillsides and open slopes
Elevation Range	3,800 to 9,300 feet
Look-Alike	Gunnison's Mariposa Lily (*Calochortus gunnisonii*)

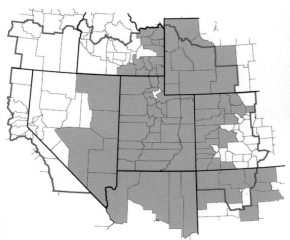

Look-Alike lavender relative

The Sego Lily is the state flower of Utah.

Feathery False Solomon's Seal
Maianthemum racemosum

Plant Size	1 to 2 feet
Flower	Close-set, small-flowered clusters (racemes) borne terminally along a stalk
	Color: White to cream Size: 2 to 4 inches long (cluster)
Bloom Time	April to June
Form/Foliage	Upright habit with large, dark green, linear, deeply veined leaves that alternate along the stems
Habitat	Shaded woodlands and moist understories
Elevation Range	4,000 to 9,900 feet
Look-Alike	Starry Solomon's Seal (*Maianthemum stellataum*)

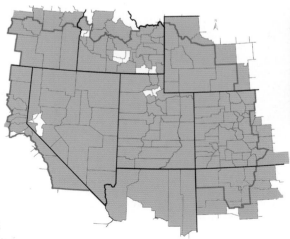

Look-Alike loose flower cluster

Referred to as "false" because it lacks
the medicinal characteristics of true
Solomon's Seal (*Polygonatum* spp.).

Elegant Death Camas
Zigadenus elegans

Plant Size	6 to 28 inches
Flower	Loose clusters of 6-petaled (tepaled) flowers, each resembling a 6-pointed star, borne along an upright stalk
	Color: White with yellow bands Size: 1/2 to 1 inch wide
Bloom Time	June to August
Form/Foliage	Upright habit with green, linear leaves (4 to 10 inches long) that are found primarily at the base of the plant
Habitat	Moist meadows and open woodlands
Elevation Range	5,200 to 13,100 feet
Look-Alike	Foothill Death Camas (*Zigadenus paniculatus*)

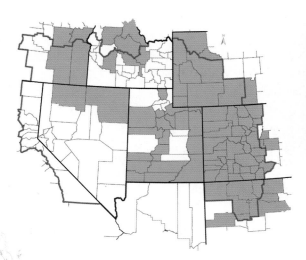

Look-Alike single stemmed

As the name implies, all death camas are poisonous to both livestock and humans. Use caution when handling this plant.

Tufted Evening-primrose
Oenothera caespitosa

Plant Size	8 to 16 inches
Flower	Large, solitary, 4-petaled, papery thin flowers borne on a drooping stalk
	Color: White, fading to pink after pollination Size: 1 1/2 to 4 inches wide
Bloom Time	April to August
Form/Foliage	Mounding habit with long (1 to 15 inch), woolly, gray-green, spear-like, lobed leaves resembling dandelion leaves
Habitat	Dry hillsides and gravelly slopes
Elevation Range	3,000 to 9,500 feet
Look-Alike	Pale Evening-primrose (*Oenothera pallida*)

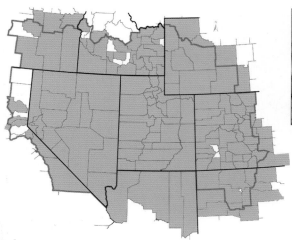

Look-Alike taller and more spindly

Oenotheras normally bloom at dusk, when pollinating nocturnal moths are most active.

White Bog Orchid

Platanthera dilatata

Plant Size	6 to 12 inches
Flower	Tiny individual flowers with a unique orchid-like shape (complete with exaggerated petals) arranged along a spike-like stalk (raceme)
	Color: White Size: Less than 1/4 inch
Bloom Time	May to August
Form/Foliage	Dainty habit with succulent, spear-like leaves that alternate along the stem
Habitat	Moist meadows, seeps or springs, and stream banks
Elevation Range	5,000 to 10,400 feet

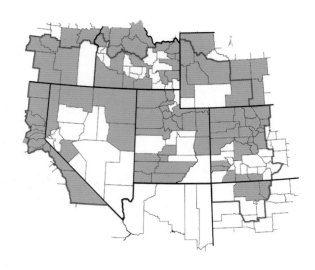

This plant is also called Scentbottle because of the "fragrant" flowers.

Flatbud Pricklypoppy
Argemone munita

Plant Size	2 to 3 feet
Flower	Loose arrangement of individual flowers consisting of 4 to 6 wrinkled petals with fuzzy yellow centers (resembling a "sunny side up" egg), borne on thick, prickly stalks
	Color: White with yellow center Size: 2 to 4 inches wide
Bloom Time	June to August
Form/Foliage	Upright habit with gray-green, extremely thorny leaves that alternate along the stems, which are also thorny
Habitat	Dry hillsides and gravelly slopes
Elevation Range	2,900 to 8,200 feet
Look-Alike	Leaves resemble those of common thistles

This plant "bleeds" a yellow or orange latex-like sap when the leaves or stems are broken.

Nuttall's Flaxflower
Leptosiphon nuttallii

Plant Size	4 to 12 inches
Flower	Loose clusters consisting of 5-petaled flowers that flare outward at right angles from a small throat
	Color: White Size: 1/4 to 3/4 inch wide
Bloom Time	June to August
Form/Foliage	Loosely mounded habit with dark, glossy green, linear leaves that are arranged in whorls of 5 to 9 around reddish stems
Habitat	Gravelly slopes and open, dry hillsides
Elevation Range	4,400 to 10,400 feet
Look-Alike	Spiny Phlox (*Phlox hoodii*) p. 28

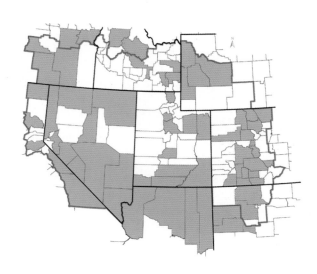

The flowers give off a slightly sweet aroma.

Spiny Phlox
Phlox hoodii

Plant Size	2 to 6 inches
Flower	Solitary tubular flowers with 5 petals that flare outward at right angles from a small throat
	Color: White to light lavender Size: 1/2 inch wide
Bloom Time	May to June
Form/Foliage	Low-growing habit with narrow, woolly, sword-like leaves that are clustered (whorled) around stiff, woody stems
Habitat	Dry hillsides and sagebrush communities
Elevation Range	4,500 to 10,400 feet
Look-Alike	Nuttall's Flaxflower (*Leptosiphon nuttallii*) p. 26

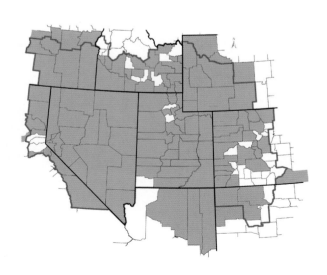

This plant is named after the famous mapmaker/artist Robert Hood.

Towering Jacob's Ladder
Polemonium foliosissimum

Plant Size   2 to 3 feet

Flower   Tight clusters (cymes) of 5-petaled flowers
                        with 5 yellow pollen-laden structures (anthers),
                        borne atop branched stalks

                        Color: White to lavender
                        Size: 1/2 to 3/4 inch long

Bloom Time   June to August

Form/Foliage   Dense, upright habit with light green, hairy
                        (pubescent), dissected leaves made up of
                        smaller leaflets (pinnately compound) that
                        resemble rungs on a ladder

Habitat   Open woodlands, moist to dry meadows, and
                        stream banks

Elevation Range   5,100 to 10,700 feet

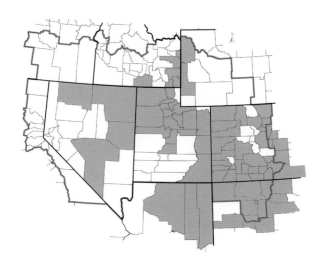

The name *foliosissimum* means "very leafy."

Whorled Buckwheat
Eriogonum heracleoides

Plant Size	8 to 14 inches
Flower	Blossoms arranged in umbrella-like clusters (umbels), borne on multiple stalks originating from a single point on stem
	Color: White to cream Size: 1 to 4 inches wide (cluster)
Bloom Time	May to September
Form/Foliage	Low-growing, mounding habit with gray-green, hairy (pubescent), oval leaves that form a loose rosette at the base of the plant
Habitat	Sandy to gravelly slopes and sagebrush communities
Elevation Range	4,300 to 10,700 feet
Look-Alike	Good luck! There are over 220 different buckwheats (*Eriogonum* spp.) in North America.

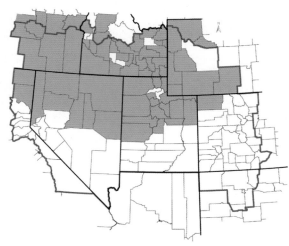

The identifying feature on this plant is
the whorl of leaves midway
along the stem.

American Bistort
Polygonum bistortoides

Plant Size	12 to 20 inches
Flower	Close-set clusters of tiny 5-petaled flowers arranged similar to a bottlebrush, borne terminally on spindly stalks (spikes)
	Color: White Size: 1 to 1 1/2 inches long (cluster)
Bloom Time	June to August
Form/Foliage	Sparsely arranged, smooth, sword-like leaves that are tapered and often tinged with rose at the base, and alternate along the stem
Habitat	Moist meadows, stream banks, and open slopes
Elevation Range	6,900 to 12,400 feet

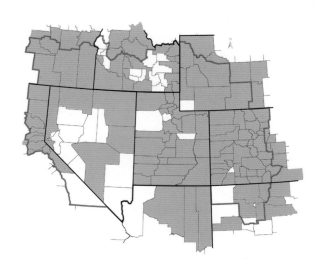

The name *bistort* comes from Latin, meaning "twice twisted," referring to the plant's twisted roots, which are a favorite treat of foraging bears.

Lanceleaf Springbeauty

Claytonia lanceolata

Plant Size	2 to 6 inches
Flower	Tiny loose clusters (racemes) of notched 5-petaled flowers borne on succulent stalks
	Color: White to pink-striped Size: 1/4 to 1/2 inch wide
Bloom Time	April to July
Form/Foliage	Dainty habit with 2 thick, succulent, opposing leaves found near the base of the plant
Habitat	Moist, shady meadows and following receding snowfields
Elevation Range	5,000 to 10,600 feet

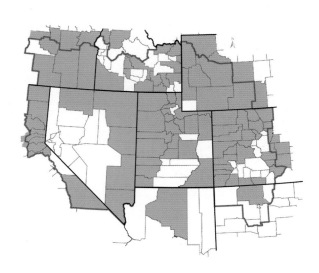

The bulb-like underground parts (corms) were eaten by Native Americans and early western settlers.

Alpine Springbeauty
Claytonia megarhiza

Plant Size	2 to 6 inches
Flower	Solitary 5-petaled flowers (with reddish veins) found along the periphery of the foliage
	Color: White Size: 1/2 to 3/4 inch wide
Bloom Time	May to August
Form/Foliage	Low-growing, tufted habit with thick, succulent, oval leaves densely arranged in a basal rosette
Habitat	Scree slopes and gravelly hillsides in subalpine to alpine sites
Elevation Range	9,900 to 11,950 feet

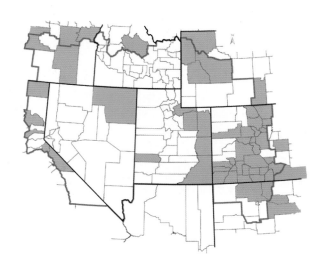

This plant grows at extremely high elevations.
Altitude sickness is a real thing...we're just saying...

White Marsh Marigold
Caltha leptosepala

Plant Size   2 to 5 inches

Flower   Solitary daisy-like (composite) flower borne
                        along a leafless stalk

                        Color: White with a yellow center
                        Size: 1/2 to 1 1/2 inches wide

Bloom Time   June to August

Form/Foliage   Small, mounding habit with heart-shaped basal
                        leaves

Habitat   Moist meadows, along stream banks, and
                        following receding snow at subalpine to alpine
                        sites

Elevation Range   7,500 to 11,500 feet

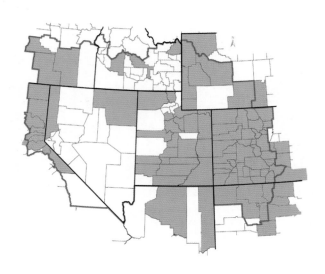

The leaves are considered toxic; use caution when handling this plant.

Woodland Strawberry
Fragaria vesca

Plant Size	6 to 12 inches
Flower	Solitary 5-petaled flowers with cone-shaped yellow centers
	Color: White with yellow center Size: 3/4 inch wide
Bloom Time	June to July
Form/Foliage	Low-growing habit with 3 broad leaflets (trifoliate) that are covered with fine hairs on the upper surface and toothed along the margins
Habitat	Moist hillsides, shady woodlands
Elevation Range	4,500 to 11,500 feet
Look-Alike	Virginia Strawberry (*Fragaria virginiana*)

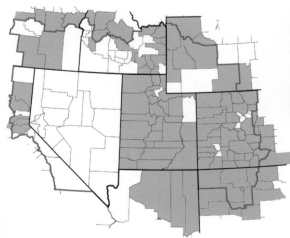

Look-Alike leaves are smooth (glabrous)

Our modern cultivars were derived from native populations in Little Cottonwood Canyon in the Wasatch Range of northern Utah.

Mat Rockspirea
Petrophytum caespitosum

Plant Size

2 to 4 inches

Flower

Dense clusters of bottlebrush-like flowers borne on single stalks (racemes) that rise above the foliage

Color: White to cream
Size: 1 to 1 1/2 inches long (cluster)

Bloom Time

August to September

Form/Foliage

Dense, mat-forming habit with small, sword-like (spatulate) leaves that form a thick basal rosette

Habitat

Exposed rocks, cracks and crevices of cliffs of limestone origin

Elevation Range

4,000 to 9,900 feet

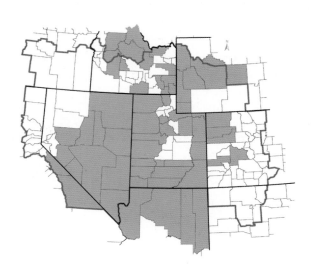

This plant was first collected by Thomas Nuttall on rocky ledges of the Rocky Mountains during his western expedition of 1834–1837.

Northern Bedstraw
Galium boreale

Plant Size   1 to 3 1/2 feet

Flower   Clusters (cymes) of tiny 4-petaled flowers borne on a stalk

Color: White
Size: About 1/4 inch wide

Bloom Time   June to August

Form/Foliage   Upright habit with narrow, sword-like leaves arranged in whorls of 4 along a thick stem

Habitat   Moist meadows and stream banks

Elevation Range 5,300 to 10,000 feet

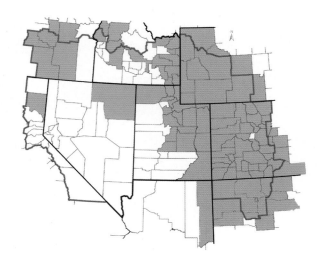

Historically, bedstraws were used to fill mattresses because the sticky stems and leaves prevented them from becoming packed down.

White Alumroot
Heuchera bracteata

Plant Size	2 to 12 inches
Flower	Tiny bell-shaped flowers tightly arranged along 1 side of an upright stalk (secund) Color: White to cream Size: Less than 1/4 inch
Bloom Time	June to August
Form/Foliage	Mounding habit with dark green, rounded leaves with scalloped edges (consisting of 5 to 7 lobes), forming a rosette at the base of the plant
Habitat	Rocky outcroppings and shady sides of cliffs
Elevation Range	5,500 to 11,500 feet

The genus *Heuchera* is named after the eighteenth-century German professor of medicine Johann Heinrich von Heucher.

Woodland Star
Lithophragma parviflorum

Plant Size   4 to 12 inches

Flower   Small clusters of deeply cleft, 5-petaled
                        flowers borne on red to purple stalks

                        Color: White, occasionally pink
                        Size: 1/2 to 3/4 inch wide

Bloom Time   April to August

Form/Foliage   Dainty habit with deeply lobed, palm-like
                        (palmately compound) leaves that are found
                        primarily at the base of the plant

Habitat   Open meadows, sagebrush communities, and
                        stream banks

Elevation Range   4,700 to 8,500 feet

Look-Alike   Fringecup Woodland Star
                        (*Lithophragma glabrum*)

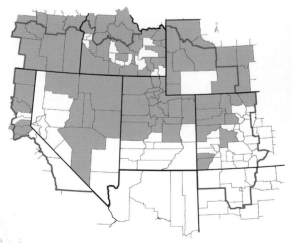

Look-Alike bulblets in
leaf axil

These are among some of the first flowers to bloom in the spring.

Side-flowered Miterwort
Mitella stauropetala

Plant Size	4 to 16 inches
Flower	Tiny 5-petaled flowers borne along a single side of an upright stalk (secund)
	Color: White Size: 1/8 to 1/4 inch wide
Bloom Time	May to August
Form/Foliage	Dainty habit with bright green, rounded and scalloped leaves found primarily at the base of the plant
Habitat	Moist meadows, shaded woodlands, and stream banks
Elevation Range	4,900 to 10,500 feet

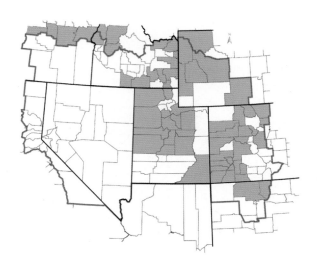

The unique flowers of miterwort have the appearance of tiny snowflakes... look close.

Spotted Saxifrage
Saxifraga bronchialis

Plant Size	1 to 5 inches
Flower	Loose, airy clusters of 5-petaled flowers with distinct spots, borne atop wiry stalks
	Color: White with purple/red spots Size: 3/8 inch wide
Bloom Time	June to September
Form/Foliage	Mounding habit with hairy (pubescent), sword-like leaves that are primarily basal, with a few alternating along reddish stems
Habitat	Moist, shaded areas and rocky crevices
Elevation Range	9,800 to 13,000 feet

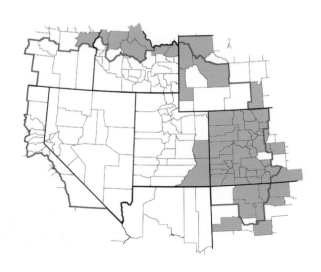

The distinct colorful spots range from yellow to orange to red to purple.

Brook Saxifrage
Saxifraga odontoloma

Plant Size	8 to 24 inches
Flower	Loose, airy clusters of tiny flowers (10 or more), borne on wiry stalks
	Color: White Size: 5 to 10 inches long (cluster)
Bloom Time	May to September
Form/Foliage	Mounding habit with bright green, deeply scalloped, fan-shaped leaves that are found at the base of the plant
Habitat	Moist meadows, slow-moving-stream banks, and bogs
Elevation Range	6,000 to 11,100 feet

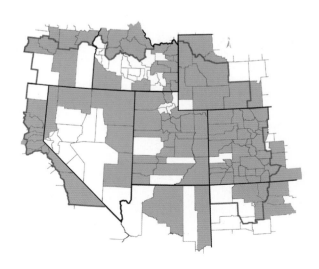

The name *Saxifraga* means "to break rock," referring to the use of the plants by herbalists to treat kidney stones.

Sickletop Lousewort
Pedicularis racemosa

Plant Size   6 to 20 inches

Flower   Clusters of irregular flowers (the upper petal is
                       uniquely sickle-shaped) arranged along a single
                       stalk (raceme)

                       Color: White to pale yellow
                       Size: 1/2 to 5/8 inch wide

Bloom Time   July to August

Form/Foliage   Loose, mounding habit with smooth, finely
                       dissected, ladder-like leaves (brown-tipped
                       when mature) that alternate along the stems

Habitat   Moist meadows and open coniferous
                       woodlands at subalpine sites

Elevation Range   7,300 to 11,000 feet

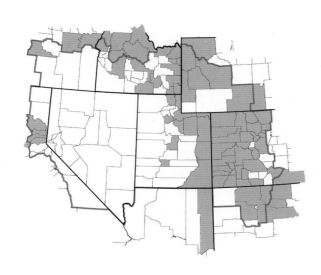

This plant is also called Parrot's Beak because of the shape of the flowers.

YELLOW FLOWERS

Owl's Claws (*Hymenoxys hoopesii*), Harney County, Oregon, 2011, Elevation: 9,700 ft.

Sprawling Spring Parsley
Cymopterus longipes

Plant Size	2 to 8 inches
Flower	Tight umbrella-like clusters (umbels) of tiny flowers borne on a short stalk
	Color: Yellow Size: 1/2 to 1 1/2 inches wide (clusters)
Bloom Time	April to May
Form/Foliage	Mounding habit with blue-green, fern-like (pinnately compound) leaves found primarily as low-growing basal rosettes
Habitat	Open hillsides and dry slopes
Elevation Range	4,400 to 5,600 feet
Look-Alike	*Cymopterus longipes* var. *ibapensis*

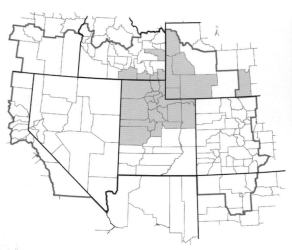

Look-Alike white flowers
with purple anthers

Sprawling Spring Parsley is one of
the earliest wildflowers to bloom in
the spring.

Mountain Dandelion
Agoseris glauca

Plant Size	4 to 30 inches
Flower	Solitary, large, daisy-like (composite) flower borne on a single stalk
	Color: Yellow Size: 1/2 to 1 1/2 inches wide
Bloom Time	May to September
Form/Foliage	Upright habit with spear-like, lobed to toothed basal leaves
Habitat	Open foothills and sagebrush communities
Elevation Range	Up to 7,800 feet
Look-Alike	Orange Mountain Dandelion (*Agoseris aurantiaca*)

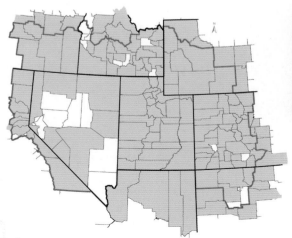

Look-Alike orange relative

Even though the common name implies that this plant associates with a notorious lower class of weeds, it is actually well behaved.

Heartleaf Arnica
Arnica cordifolia

Plant Size	8 inches to 1 foot
Flower	Solitary daisy-like (composite) flowers with 8 to 13 outer petals (ray flowers), borne on a single stalk
	Color: Yellow Size: 2 to 3 inches wide
Bloom Time	May to August
Form/Foliage	Woolly, heart-shaped leaves that are found opposite each other on low, ground-hugging stems
Habitat	Open woodlands, stream banks, aspen and spruce groves
Elevation Range	5,100 to 11,500 feet
Look-Alike	Other arnicas (*Arnica* spp.)

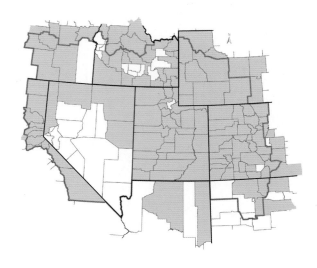

The name *cordifolia* refers to the heart-shaped leaves.

Largeleaf Balsamroot
Balsamorhiza macrophylla

Plant Size	12 to 40 inches
Flower	Solitary daisy-like (composite) flowers that arise from a single long stalk
	Color: Yellow Size: 1 1/2 to 2 1/2 inches wide
Bloom Time	May to July
Form/Foliage	Loosely mounding habit with large (12 to 24 inches), green, deeply lobed, hairy (pubescent) leaves that are found at the base of the plant
Habitat	Mountain slopes and sagebrush communities
Elevation Range	5,000 to 7,500 feet
Look-Alike	Hooker's Balsamroot (*Balsamorhiza hookeri*)

Look-Alike sandpapery leaves

Look for variations of balsamroots, as they can readily hybridize with other species.

Arrowleaf Balsamroot
Balsamorhiza sagittata

Plant Size	1 to 2 feet
Flower	Solitary daisy-like (composite) flowers that arise from a single long stalk
	Color: Yellow Size: 2 to 4 inches wide
Bloom Time	May to June
Form/Foliage	Loosely upright habit with large, woolly (pubescent), gray-green, distinctly arrow-shaped leaves (up to 1 foot long) that are found at the base of the plant
Habitat	Mountain slopes and sagebrush communities
Elevation Range	4,000 to 9,000 feet
Look-Alike	Mule's Ear (*Wyethia amplexicaulis*) p. 94

The thick taproots were used as a food source by Native Americans and early western settlers.

Tapertip Hawksbeard
Crepis acuminata

Plant Size	8 to 28 inches
Flower	Loose clusters of numerous (30 to 100) daisy-like (composite) flowers, each petal having finely toothed tips
	Color: Yellow Size: 3/4 inch wide
Bloom Time	May to August
Form/Foliage	Upright habit with long gray-green, spear-like, deeply lobed leaves found primarily at the base of the plant
Habitat	Dry hillsides and sagebrush communities
Elevation Range	4,600 to 9,500 feet

We have no
idea why this
plant is called
Hawksbeard...birds
don't have facial
hair.

Rocky Mountain Dwarf Sunflower
Helianthella uniflora

Plant Size	1 1/2 to 3 feet
Flower	Solitary daisy-like (composite) flowers having 13 to 21 outer petals (ray flowers), borne on a single stalk
	Color: Yellow
	Size: 1 1/2 to 2 1/2 inches wide
Bloom Time	June to August
Form/Foliage	Upright habit with spear-like leaves (2 inches wide and 6 inches long) that are rough to the touch and found opposite each other on the stems
Habitat	Open woodlands, dry hillsides, and lower canyon slopes
Elevation Range	4,500 to 10,300 feet
Look-Alike	5-nerved Sunflower (*Helianthella quinquenervis*)

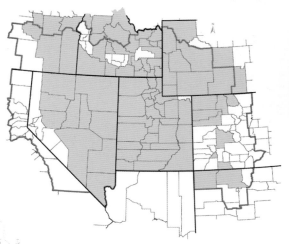

Look-Alike taller, with larger leaves and flowers

The seeds were eaten by
Native Americans.

Showy Goldeneye
Heliomeris multiflora

Plant Size	1 to 3 feet
Flower	Multiple daisy-like (composites) flowers with 10 to 16 outer petals (ray flowers), borne along a single stalk
	Color: Yellow with dark golden centers (disc flowers) Size: 1/2 to 1 inch wide
Bloom Time	June to September
Form/Foliage	Upright, airy habit with sparse spear-like leaves that curl slightly upright and are found opposite each other on the stems
Habitat	Dry hillsides, open woodlands, and disturbed sites
Elevation Range	4,500 to 10,100 feet
Look-Alike	Rocky Mountain Dwarf Sunflower (*Helianthella uniflora*) p. 74

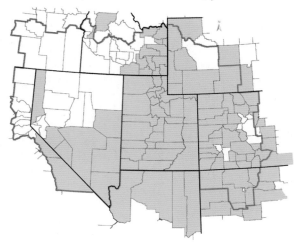

Helios in Greek means "sun"
and *meris* means "part of."

Tundra Hymenoxys
Hymenoxys grandiflora

Plant Size	4 to 12 inches
Flower	Large, solitary, daisy-like (composite) flowers borne terminally atop a stalk
	Color: Yellow Size: 1 to 2 1/2 inches wide
Bloom Time	July to August
Form/Foliage	Low, mounding habit with narrow, deeply dissected, hairy (pubescent) leaves that alternate along the stems
Habitat	Gravelly slopes, open meadows, and rocky sites
Elevation Range	10,000 to 12,200 feet

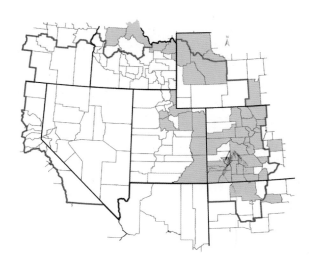

Tundra Hymenoxys is able to retain heat in alpine conditions because of the fine hairs that cover its stems and leaves.

Owl's Claws
Hymenoxys hoopesii

Plant Size	1/2 to 3 feet
Flower	Multiple daisy-like (composite) flowers with cleft-tipped outer petals (ray flowers), arranged in loose clusters atop smooth stalks
	Color: Yellow
	Size: 1 3/4 to 3 1/4 inches wide
Bloom Time	June to August
Form/Foliage	Loosely upright habit with long, smooth, light green leaves found primarily at the base of the plant
Habitat	Moist meadows, open woodlands, stream banks, and slopes
Elevation Range	6,200 to 11,700 feet
Look-Alike	Hairy Arnica (*Arnica mollis*)

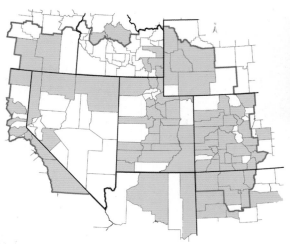

Look-Alike leaves are hairy (pubescent)

Historically, Native Americans sniffed the dried petals to induce sneezing in order to clear their sinuses, giving it the common name Sneezeweed.

Cutleaf Coneflower
Rudbeckia laciniata

Plant Size	1 1/2 to 6 1/2 feet
Flower	Large, daisy-like (composite) flowers with 6 to 16 drooping outer petals (ray flowers) radiating from a central dome-shaped cone (disc flowers)
	Color: Yellow Size: 3 to 5 inches wide
Bloom Time	July to September
Form/Foliage	Upright habit with deeply dissected 3 to 5 lobed leaves alternating along tall stems
Habitat	Moist meadows, aspen groves, and along stream banks
Elevation Range	6,200 to 7,500 feet
Look-Alike	Black-eyed Susan (*Rudbeckia hirta*)

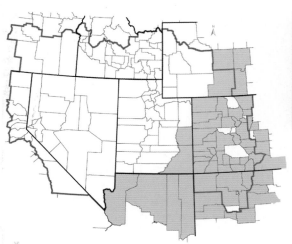

Look-Alike dark
central disk

Cultivated varieties of this plant can be found in the nursery trade.

Lambstongue Groundsel
Senecio integerrimus

Plant Size	18 to 20 inches
Flower	Multiple clusters of individual daisy-like flowers (composites), borne along single unbranched stalks
	Color: Yellow Size: 3/4 inch wide
Bloom Time	May to August
Form/Foliage	Thick, spear-like leaves (narrow and toothed) that are alternate and found primarily at the base of the plant
Habitat	Dry to moderately wet soils and disturbed sites
Elevation Range	4,500 to 11,000 feet
Look-Alike	Other groundsels (*Senecio* spp.)

This is one of the earlier blooming groundsels, also called Butterweed.

Tall Groundsel
Senecio serra

Plant Size	2 to 6 feet
Flower	Multiple clusters of individual daisy-like flowers (composites), borne along single branched stalks
	Color: Yellow with black-tipped bracts
	Size: 1/2 inch wide
Bloom Time	June to August
Form/Foliage	Upright habit with long, spear-like, serrated leaves that alternate along the stems
Habitat	Dry to moderately wet soils and disturbed sites
Elevation Range	6,000 to 11,000 feet
Look-Alike	Other groundsels (*Senecio* spp.)

There are over 25 different groundsels (species) throughout the Mountain West.

Arrowleaf Groundsel
Senecio triangularis

Plant Size	1 to 5 feet
Flower	Multiple clusters of individual daisy-like flowers (composites), borne along single branched stalks
	Color: Yellow Size: 1/2 to 1 inch wide
Bloom Time	June to August
Form/Foliage	Upright habit with large, arrow-shaped leaves that alternate along the stems
Habitat	Open woodlands and meadows
Elevation Range	6,000 to 11,000 feet
Look-Alike	Other groundsels (*Senecio* spp.)

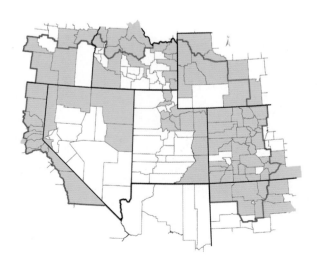

The name *triangularis* describes the distinct triangular shape of the leaves.

Canada Goldenrod
Solidago canadensis

Plant Size	2 1/2 to 4 feet
Flower	Feathery clusters of individual daisy-like (composite) flowers, borne along a single side of the stalk (secund) Color: Yellow Size: 1/10 to 1/4 inch wide
Bloom Time	July to October
Form/Foliage	Upright habit with narrow, spear-like, serrated leaves that alternate along the stems
Habitat	Open meadows, stream banks, and dry, open sites
Elevation Range	6,500 to 10,100 feet
Look-Alike	Baby Goldenrod (*Solidago nana*)

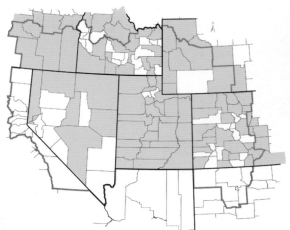

Look-Alike smaller, with rounded leaves

The genus name *Solidago* means "to make whole," which refers to the described medicinal properties of goldenrods.

Stemless Mock Goldenweed
Stenotus acaulis

Plant Size	4 to 6 inches
Flower	Solitary daisy-like (composite) flowers with 6 to 15 outer petals (ray flowers), borne on a single stalk
	Color: Yellow
	Size: 1/2 to 3/4 inch wide
Bloom Time	May to June
Form/Foliage	Mounding habit with gray-green, sword-like, alternate leaves found primarily at the base of the plant
Habitat	Dry, rocky slopes of high elevations
Elevation Range	4,500 to 10,400 feet
Look-Alike	Woolly Sunflower (*Eriophyllum lanatum*)

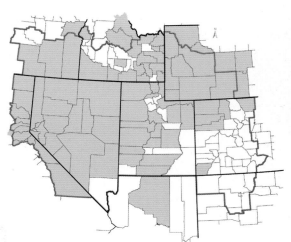

Look-Alike leaves and stems are covered with white hairs

The name *acaulis* describes the
petiole-less (sessile) leaves.

Mule's Ear
Wyethia amplexicaulis

Plant Size	2 to 3 feet
Flower	Solitary, large, daisy-like (composite) flowers, borne on a single long stalk
	Color: Yellow, occasionally white Size: 3 to 4 inches wide
Bloom Time	May to June
Form/Foliage	Loosely upright habit with large, dark green, shiny (totally hairless) leaves found primarily at the base of the plant
Habitat	Gravelly foothills and open meadows
Elevation Range	4,600 to 9,500 feet
Look-Alike	✓ Arrowleaf Balsamroot (*Balsamorhiza sagittata*) p. 70

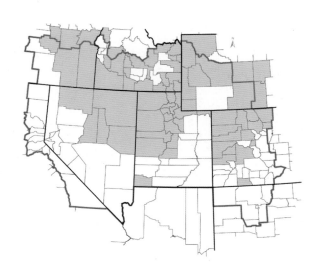

The name *amplexicaulis* refers to the stem-clasping, petiole-less (sessile) leaves.

Puccoon
Lithospermum ruderale

Plant Size	8 to 28 inches
Flower	Clusters of 5-petaled flowers, borne atop a stalk, occasionally in the leaf axis
	Color: Yellow Size: 1/5 to 1/3 inch wide
Bloom Time	April to June
Form/Foliage	Upright habit with gray-green, hairy (pubescent), sword-like leaves that alternate along the stems
Habitat	Hillsides, open slopes, and sagebrush communities
Elevation Range	4,300 to 8,800 feet

This plant is also called "Stone-seed" because each flower produces four small nutlets that resemble small pebbles.

Western Wallflower
Erysimum capitatum

Plant Size	18 to 24 inches
Flower	Multiple clusters of cross-shaped flowers, each with 4 petals, borne on a single stalk
	Color: Yellow Size: 1/2 to 3/4 inch wide
Bloom Time	May to June
Form/Foliage	Airy habit with extremely narrow, sword-like, finely toothed leaves that alternate along the stem
Habitat	Open rocky slopes from valleys to high elevations
Elevation Range	2,600 to 12,000 feet
Look-Alike	Other members of the mustard family, all of which have "cross-shaped" flowers with 4 petals.

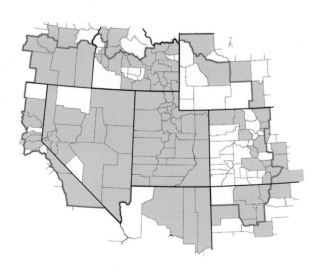

Historically, Native Americans ground
the plant, mixed it with water, and
applied it as a sun block.

Prince's Plume
Stanleya pinnata

Plant Size	1 1/2 to 5 feet
Flower	Elongated clusters (racemes) of 4-petaled flowers with distinct long sprays of yellow, projecting, whisker-like flower parts (stamens)
	Color: Yellow Size: 3/8 to 5/8 inch long
Bloom Time	July to September
Form/Foliage	Tall, loose habit with deeply lobed leaves that alternate along the stem
Habitat	Dry hillsides, sagebrush communities, and disturbed slopes
Elevation Range	3,000 to 9,100 feet
Look-Alike	*Stanleya pinnata* var. *integrifolia*

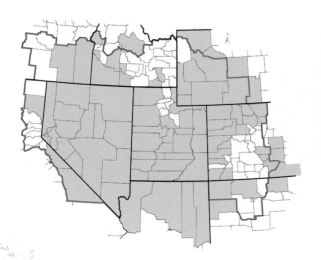

Look-Alike entire, spearlike leaves

This plant accumulates selenium in its foliage, which can be toxic to livestock.

Spearleaf Stonecrop
Sedum lanceolatum

Plant Size	2 to 6 inches
Flower	Multiple clusters of 5-petaled star-shaped flowers, borne on a single stalk (cyme) and resembling stars
	Color: Yellow
	Size: 1/4 to 1/3 inch wide
Bloom Time	June to August
Form/Foliage	Low-growing habit with small, gray-green, oval (pointed tip), succulent leaves found primarily at the base of the plant
Habitat	Open rocky slopes and shallow gravelly sites
Elevation Range	5,000 to 12,000 feet

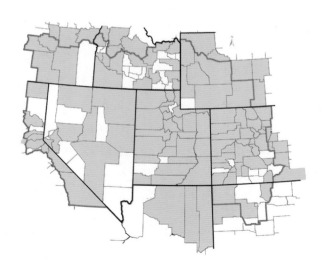

Stonecrops have thick, succulent leaves to help store water during long periods of drought.

Mountain Golden Pea
Thermopsis rhombifolia

Plant Size	6 to 28 inches
Flower	Loose clusters of pea-like flowers borne along upright stalks (racemes)
	Color: Yellow Size: 1/2 to 1 inch wide
Bloom Time	May to August
Form/Foliage	Upright habit with narrow, clover-like (trifoliate) leaves that alternate along the stems
Habitat	Open meadows and woodlands
Elevation Range	4,400 to 7,500 feet
Look-Alike	Spreadfruit Golden Pea (*Thermopsis divaricarpa*)

Look-Alike broader leaflets. Primarily found in drier soils in the Central Rockies of Colorado.

The seed pods of
this plant resemble
hairy garden peas
that point upward.

Yellow Avalanche Lily
Erythronium grandiflorum

Plant Size	4 to 12 inches
Flower	1 to several drooping (pendulous), fragrant flowers whose petals, when mature, turn upward (recurve); flowers are borne terminally on smooth stalks
	Color: Yellow Size: 1 to 1 1/2 inches wide
Bloom Time	April to July
Form/Foliage	Dainty form with 2 flattened, oval (pointed tip) leaves found primarily at the base of the plant
Habitat	Sparsely wooded hillsides and following receding snowfields
Elevation Range	4,600 to 10,400 feet
Look-Alike	Yellow Fritillary (*Fritillaria pudica*) p. 108

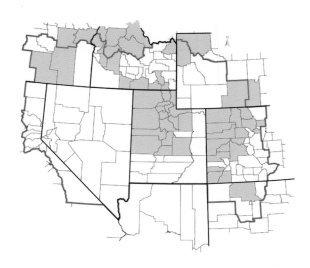

In some locations, the flowers can be so abundant that they cover hillsides like a colorful blanket.

Yellow Fritillary
Fritillaria pudica

Plant Size	6 to 10 inches
Flower	Nodding (pendulous), 6-petaled (tepaled) flower, borne terminally on a single stalk
	Color: Yellow to orange with red band
	Size: 3/4 inch long
Bloom Time	May to June
Form/Foliage	Dainty habit with bright green, sword-like, twisted leaves that alternate along the stem
Habitat	Open slopes, woodlands, meadows, sagebrush communities, and following receding snowfields
Elevation Range	5,000 to 8,400 feet
Look-Alike	Yellow Avalanche Lily (*Erythronium grandiflorum*) p. 106

Flowers give way to a three-sectioned fruit that is puffy and resembles a marshmallow.

Blazingstar
Mentzelia laevicaulis

Plant Size	1 1/2 to 2 1/2 feet
Flower	Solitary flowers consisting of 5 narrow, flattened petals, with distinct sprays of whisker-like flower parts (pistils and stamens), borne atop thick stalks
	Color: Yellow, orange to white Size: 1 to 3 inches wide
Bloom Time	April to August
Form/Foliage	Upright habit with green, spear-like, toothed leaves that alternate along the stems, are covered with stiff hairs, and feel like sandpaper
Habitat	Gravelly hillsides, dry slopes, and disturbed sites
Elevation Range	2,600 to 7,900 feet

The flowers open in the evening and close during part of the day.

Sulphur Flower Buckwheat
Eriogonum umbellatum

Plant Size	12 to 14 inches
Flower	Blossoms arranged in umbrella-like clusters (umbels), borne on multiple stalks that originate from a single point or sturdy stalk
	Color: Yellow, white, and occasionally red Size: 1 to 4 inches (clusters)
Bloom Time	June to September
Form/Foliage	Mounding habit with green, oval leaves that are smooth above and hairy (pubescent) below and form a loose rosette at the base of the plant
Habitat	Sandy to gravelly slopes, open woodlands, and sagebrush communities
Elevation Range	6,900 to 12,400 feet
Look-Alike	There are 41 described varieties of this species alone that vary in plant size, flower color, and bloom time

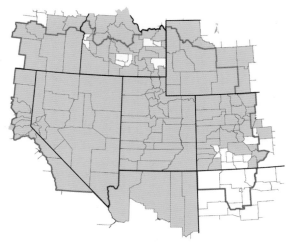

The genus name *Eriogonum* means "woolly leg," referring to the hairy stems.

Hillside Buttercup
Ranunculus jovis

Plant Size	2 to 4 inches
Flower	Solitary waxy flowers, each with 3 to 7 petals (sepals), borne 1 to 4 per stalk
	Color: Yellow Size: 1/2 to 1 inch wide
Bloom Time	April to July
Form/Foliage	Low-growing habit with succulent, deeply lobed, narrow, 3-parted (trifoliate) leaves found opposite each other on the stems
Habitat	Dry, open slopes, sagebrush communities, and by receding snowfields
Elevation Range	5,400 to 9,000 feet
Look-Alike	Alpine Buttercup (*Ranunculus adoneus*)

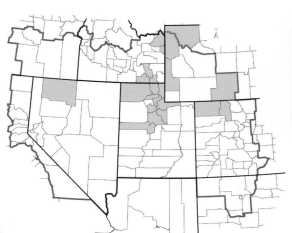

Look-Alike found at higher elevations

There are many different buttercups found in the Mountain West.

Alpine Avens
Geum rossii

Plant Size	2 to 12 inches
Flower	Loose clusters (cymes) of 1 to 4 flowers with 5 petals, borne on erect stalks
	Color: Yellow Size: 3/8 to 3/4 inch wide
Bloom Time	July to August
Form/Foliage	Low-growing, tufted habit with fern-like leaves found primarily at the base of the plant
Habitat	Scree slopes and subalpine to alpine meadows
Elevation Range	9,150 to 12,900 feet
Look-Alike	Cinquefoils (*Potentilla* spp.) p. 120

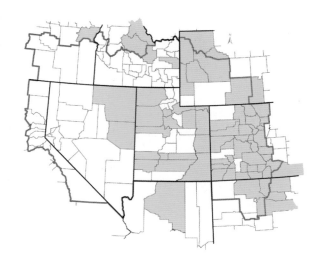

Also called Ross's Avens in honor of Arctic explorer James Ross.

Gordon's Ivesia
Ivesia gordonii

Plant Size	1 to 2 feet
Flower	Tight globe-like clusters (cymes) of star-shaped flowers, borne on naked stalks
	Color: Yellow Size: 3/4 to 1 1/2 inches wide (clusters)
Bloom Time	June to August
Form/Foliage	Low-growing, tufted habit with thin, fern-like leaves, resembling thick feathers, found at the base of the plant
Habitat	Alpine tundra, sagebrush communities, and high, rocky slopes
Elevation Range	6,600 to 10,900 feet
Look-Alike	Buckwheats (*Eriogonum* spp.) pp. 32, 112

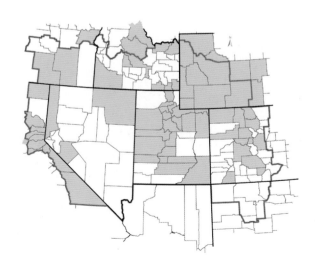

These plants can
thrive in very
shallow soil.

Sticky Cinquefoil
Potentilla glandulosa

Plant Size	4 to 16 inches
Flower	Loose clusters (2 to 3) of 5-petaled flowers borne on erect stalks
	Color: Yellow Size: 3/4 inch wide
Bloom Time	June to August
Form/Foliage	Loosely mounded habit with sticky leaves, made up of smaller leaflets (pinnately compound), found primarily at the base of the plant
Habitat	Open woodlands, mountain meadows, and alpine tundra
Elevation Range	4,600 to 10,400 feet
Look-Alike	Slender Cinquefoil (*Potentilla gracilis*)

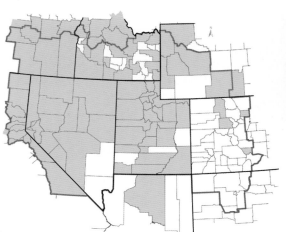

Look-Alike palmate leaves

Potentillas are
found on every
continent except
Antarctica.

Sulphur Paintbrush
Castilleja sulphurea

Plant Size	10 to 22 inches
Flower	The flowers are inconspicuous, tubular-shaped, and usually green to yellow, while the noted bright colors are actually modified leaves (bracts) Color: Yellow to creamy white Size: 3/4 to 1 1/8 inches long
Bloom Time	June to August
Form/Foliage	Upright habit with entire, narrow, spear-like leaves that are smooth (glabrous), have a toothed tip, and alternate along the stem
Habitat	Open woodlands and exposed slopes in subalpine areas
Elevation Range	7,800 to 11,200 feet
Look-Alike	Western Paintbrush (*Castilleja occidentalis*)

Look-Alike shorter, with hairy leaves

Sulphur Paintbrush is often called Yellow Paintbrush; however, we have observed it with more of a creamy white color.

Seep Monkeyflower
Mimulus guttatus

Plant Size	1 to 2 feet
Flower	Unique tubular, face-like (bilaterally symmetrical) flowers borne in pairs in the axils of the leaves
	Color: Yellow with brown or red spots in the throat Size: 1/2 to 1 inch wide
Bloom Time	May to September
Form/Foliage	Upright habit with dark green, oval, toothed leaves found opposite each other on the stems
Habitat	Wet meadows, bogs, and stream banks
Elevation Range	3,000 to 10,300 feet
Look-Alike	Lewis's Monkeyflower (*Mimulus lewisii*) p. 200

The name is derived from the Latin word *mimulus*, meaning "comic actor." Supposedly the flower has the likeness of a mischievous monkey's face.

Tolmie's Owlclover
Orthocarpus tolmiei

Plant Size	4 to 12 inches
Flower	The flowers are inconspicuous and tubular-shaped, while the noted bright colors are actually modified leaves (bracts) Color: Yellow Size: 1/3 to 1/2 inch long
Bloom Time	July to September
Form/Foliage	Dainty habit with slender, gray-green, sword-like leaves that alternate along hairy (pubescent) stems
Habitat	Open woodlands, dry meadows, and hillsides
Elevation Range	5,600 to 10,400
Look-Alike	Sulphur Paintbrush (*Castilleja sulphurea*) p. 122

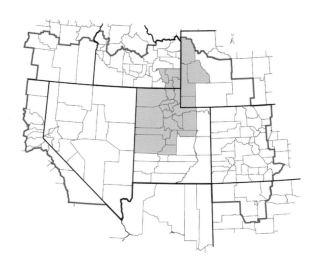

All owlclovers are annuals, completing their life cycle in one season.

Goosefoot Yellow Violet
Viola purpurea

Plant Size	2 to 8 inches
Flower	Solitary, 5-petaled flowers (consisting of 3 basal petals and 2 upward-pointing petals), borne in the leaf axils
	Color: Yellow with brown stripes in the flower's throat Size: 1/4 to 1/2 inch wide
Bloom Time	April to August
Form/Foliage	Low habit with dark green, oval leaves that are rough like sandpaper and form a loose basal rosette
Habitat	Gravelly slopes and sagebrush communities
Elevation Range	5,000 to 9,900 feet
Look-Alike	Valley Violet (*Viola vallicola*)

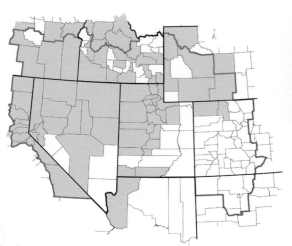

Look-Alike bright green spear-like leaves

The leaves resemble the pad of a goose's foot.

ORANGE FLOWERS

Gooseberryleaf Globemallow (*Sphaeralcea grossulariifolia*), Box Elder County, Utah, 2010, Elevation: 4,505 ft.

Indian Blanketflower
Gaillardia aristata

Plant Size	1 to 3 feet
Flower	Solitary daisy-like (composite) flowers with 12 outer petals (ray flowers), each with 3 distinct lobes, borne on a single stalk
	Color: Orange to yellow with brownish red center
	Size: 2 to 3 inches wide
Bloom Time	June to September
Form/Foliage	Upright, loose habit with green, oval (pointed tip), hairy (pubescent) leaves that alternate along the stems
Habitat	Open meadows, sagebrush communities, and disturbed sites
Elevation Range	5,000 to 9,400 feet

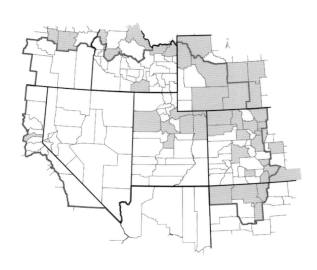

The name *aristata* means "bristly," and refers to the texture of the brown disc flowers.

Gooseberryleaf Globemallow
Sphaeralcea grossulariifolia

Plant Size	6 to 20 inches
Flower	Solitary flowers consisting of 5 vibrant petals, radiating along a central stalk
	Color: Orange Size: 3/4 to 1 1/2 inches wide
Bloom Time	June to August
Form/Foliage	Upright habit with deeply lobed (3 to 5), gray-green, hairy (pubescent) leaves that are variable in shape and size
Habitat	Deserts, sagebrush communities, and dry hillsides
Elevation Range	2,600 to 7,000 feet

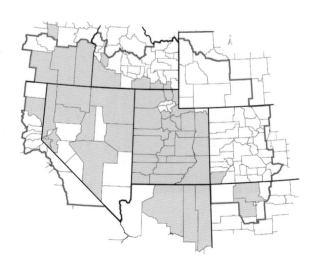

The leaves resemble those of a currant, or gooseberry.

Large-flowered Collomia
Collomia grandiflora

Plant Size	8 to 40 inches
Flower	Loose, dome-shaped clusters consisting of many 5-petaled, trumpet-shaped flowers
	Color: Orange to peach Size: 3/4 to 1 1/4 inches long
Bloom Time	May to August
Form/Foliage	Upright habit with gray-green, sword-like leaves (2 to 4 inches long) that alternate along the stems
Habitat	Dry hillsides, open woodlands, and disturbed sites
Elevation Range	4,700 to 8,000 feet

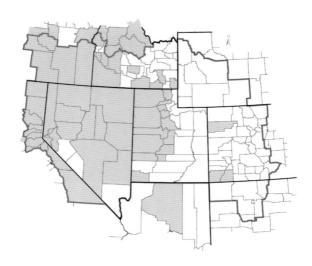

The orange petals and blue anthers demonstrate one of nature's great color combinations.

Narrowleaf Paintbrush
Castilleja linariifolia

Plant Size	1 to 1 1/2 feet
Flower	The flowers are inconspicuous, tubular-shaped, and usually green to yellow, while the noted bright colors are actually modified leaves (bracts) Color: Orange to red Size: 1 to 1 3/4 inches long
Bloom Time	July to August
Form/Foliage	Upright habit with narrow linear leaves that are mostly smooth (with some fine hairs) and alternate along the stem
Habitat	Dry hillsides, sagebrush communities, and coniferous forests
Elevation Range	3,700 to 10,950 feet

This is the state flower of Wyoming.

RED FLOWERS

Early Paintbrush (*Castilleja chromosa*), Bannock County, Idaho, 2010
Elevation: 4,670 ft.

King's Crown
Rhodiola rosea

Plant Size	2 to 8 inches
Flower	A tight, flat-topped cluster (cyme) of individual star-shaped flowers that form a dense head that is borne atop a thick, succulent stalk
	Color: Red Size: 1/8 to 1/2 inch long
Bloom Time	July to August
Form/Foliage	Loose, low-growing habit with thick, oval (pointed tip), succulent, upward-turning leaves that alternate along erect stems
Habitat	Moist meadows, stream banks, and seeps in rocks
Elevation Range	8,900 to 10,950 feet
Look-Alike	Queen's Crown (*Rhodiola rhodantha*)

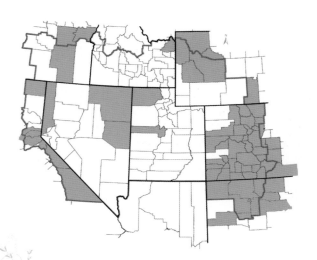

Look-Alike loose, pink flower clusters

Also called Roseroot because its thick roots supposedly smell like roses. Just trust the authors and put the spade down.

Garrett's Firechalice
Epilobium canum

Plant Size	12 to 18 inches
Flower	Tubular, trumpet-shaped flowers with long flower parts (pistils and stamens) that protrude from the flower's throat
	Color: Red to orange-red Size: 1 to 1 1/2 inches long
Bloom Time	September to October
Form/Foliage	Mounding habit with light green, oval (pointed tip), woolly leaves, each with a prominent midvein, alternating along the stems
Habitat	Rocky outcroppings, dry hillsides, and disturbed sites
Elevation Range	5,000 to 9,900 feet

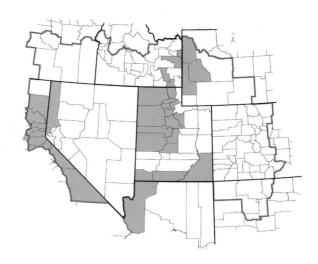

The bright, tubular flowers are pollinated by hummingbirds.

The Mountain West forms include subspecies *garrettii* and *latifolium*.

Scarlet Gilia
Ipomopsis aggregata

Plant Size	8 to 40 inches
Flower	Loose clusters of fused, 5-petaled, trumpet-shaped flowers (with flared tips, creating a star shape) alternating along a central stalk
	Color: Red, pink, or white with red to yellow spots Size: 1/2 to 1 1/2 inches long
Bloom Time	May to September
Form/Foliage	Airy habit with silvery-green, fern-like, alternating leaves (pinnately compound) that are found primarily at the base of the plant
Habitat	Dry meadows, open woodlands, and rocky slopes
Elevation Range	4,500 to 10,400 feet

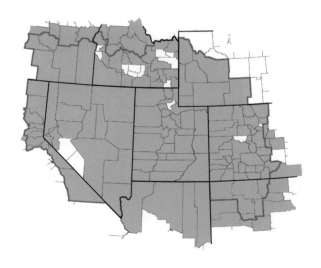

The crushed flowers are said to smell like a skunk's spray...back away slowly.

Western Columbine
Aquilegia formosa

Plant Size	1 to 3 feet
Flower	Intricate drooping (pendulous) flowers consisting of 5 petals (spurs) and 5 alternating sepals, which are flattened or flared, borne atop a single stalk
	Color: Red Size: 3/8 to 1 inch wide
Bloom Time	April to August
Form/Foliage	Airy habit with bright green, thin leaves that are distinctly 3-lobed (trifoliate) and found primarily at the base of the plant
Habitat	Shady woodlands and wooded understories
Elevation Range	4,400 to 9,950 feet
Look-Alike	Yellow Columbine (*Aquilegia flavescens*)

Look-Alike yellow
flowers with
shorter spurs

Columbine originates from the Latin root *columbina,* which means "dove-like."

Early Paintbrush
Castilleja chromosa

Plant Size	6 to 14 inches
Flower	The flowers are inconspicuous, tubular-shaped, and usually green to yellow, while the noted bright colors are actually modified leaves (bracts)

Color: Brilliant red to orange
Size: 3/4 to 1 1/4 inches long |
Bloom Time	May to June
Form/Foliage	Upright habit (seldom branched) with narrow, deeply lobed (3 to 5), spear-like leaves that are covered with fine hair (pubescence) and alternate along the stems
Habitat	Dry slopes, open woodlands, and sagebrush communities
Elevation Range	3,200 to 8,200 feet
Look-Alike	Other paintbrushes (*Castilleja* spp.)

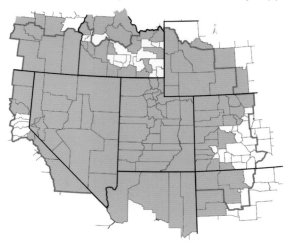

Early Paintbrush is one of the most common and earliest blooming paintbrushes along the Wasatch Range.

Scarlet Paintbrush
Castilleja miniata

Plant Size	6 to 14 inches
Flower	The flowers are inconspicuous, tubular-shaped, and usually green to yellow, while the noted bright colors are actually modified leaves (bracts) Color: Red to crimson Size: 1 to 1 3/4 inches long
Bloom Time	July to August
Form/Foliage	Upright habit (with branched heads) with entire, smooth (glabrous), spear-like leaves that alternate along the stems
Habitat	Meadows and open woodlands of alpine and subalpine locations
Elevation Range	8,200 to 11,500 feet
Look-Alike	Rosy Paintbrush (*Castilleja rhexiifolia*)

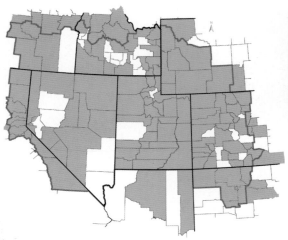

Look-Alike found at
high elevations and has
unbranched heads

Paintbrushes are usually connected
(at their roots) to either sagebrush
or grasses (hemiparasitic), making it
difficult to cultivate or grow them.

Firecracker Beardtongue
Penstemon eatonii

Plant Size	1 1/2 to 3 1/2 feet
Flower	4 to 12 tiers consisting of 1 to 2 tubular flowers per tier (cyme), borne along 1 side of a stalk (secund)
	Color: Red Size: 1 to 1 1/4 inches long
Bloom Time	June to July
Form/Foliage	Sparse, upright habit with thick, oval, evergreen leaves that are found primarily at the base of the plant, with a select few being opposite along the stem
Habitat	Dry slopes, sagebrush communities, and coniferous woodlands
Elevation Range	2,750 to 11,150 feet
Look-Alike	Beardlip Beardtongue (*Penstemon barbatus*)

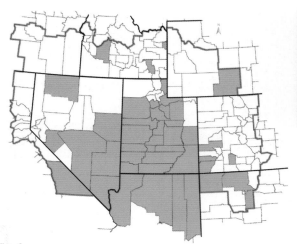

Look-Alike 2 to 4 flowers extending away from the stem on a forked, elongated stalk (pedicel)

This beardtongue is pollinated by hummingbirds.

PINK FLOWERS

**Parry's Primrose (*Primula parryi*), Salt Lake County, Utah, 2010
Elevation: 9,850 ft.**

Rosy Pussytoes
Antennaria rosea

Plant Size	8 to 16 inches tall
Flower	Tight, dome-shaped clusters of flowers (composite) resembling a fuzzy cotton swab
	Color: Pink Size: 1/10 to 1/4 inch long
Bloom Time	June to August
Form/Foliage	Low, mounding habit with gray-green, fuzzy (pubescent) leaves that are sword-like and primarily basal
Habitat	Open woodlands, dry meadows, and rocky slopes
Elevation Range	5,000 to 11,890 feet
Look-Alike	Littleleaf Pussytoes (*Antennaria microphylla*)

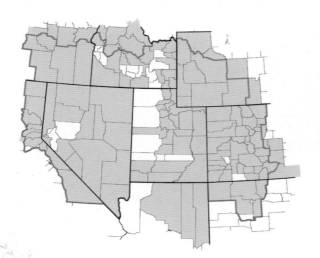

Look-Alike
predominantly white

Individual plants usually have either male or female flowers.

Twinflower
Linnaea borealis

Plant Size	2 to 4 inches
Flower	2 identical drooping (pendulous), bell-shaped flowers arranged atop a forked stalk
	Color: Pink Size: 1/4 to 5/8 inch long
Bloom Time	June to September
Form/Foliage	Low, creeping habit with rounded, slightly toothed evergreen leaves that are found opposite each other
Habitat	Moist, shady areas and wooded understories
Elevation Range	5,900 to 9,450 feet

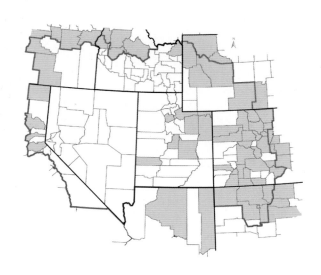

The genus name *Linnaea* is in honor of the early Swedish botanist Carolus Linnaeus. Linnaeus developed the system of scientific naming of plants and animals.

Moss Campion
Silene acaulis

Plant Size	4 to 6 inches
Flower	Solitary tubular flowers with 5 notched petals that flare outward at right angles from the small throat
	Color: Pink Size: 1/4 to 3/8 inch wide
Bloom Time	July to August
Form/Foliage	Low-growing habit forming a cushion, with narrow, sword-like leaves that are opposite and densely arranged
Habitat	Rocky slopes and gravelly hillsides in subalpine to tundra sites
Elevation Range	8,700 to 12,900 feet
Look-Alike	Longleaf Phlox (*Phlox longifolia*) p. 186

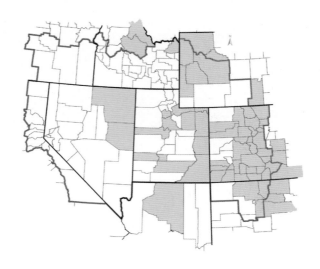

The flowers have
a distinct "spicy"
aroma.

Rocky Mountain Beeplant
Cleome serrulata

Plant Size	12 to 60 inches
Flower	Large clusters of 4-petaled flowers with distinct long sprays of green, projecting, whisker-like flower parts (stamens)
	Color: Pink to purple Size: 2 to 3 inches wide
Bloom Time	June to October
Form/Foliage	Tall, upright habit with clover-like (trifoliate), deeply divided leaves that alternate along the stems
Habitat	Open plains, disturbed sites, and sagebrush communities
Elevation Range	2,300 to 9,000 feet

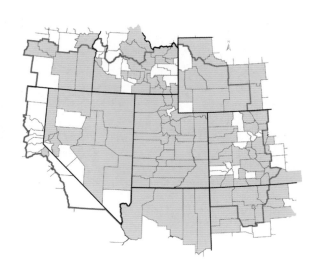

Because of the rich nectar produced by cleomes, they are a favorite of pollinating bees.

Utah Milkvetch
Astragalus utahensis

Plant Size	6 to 8 inches
Flower	Clusters of pea-like flowers, 2 to 8 per cluster at the end of a single stalk (raceme), resembling large folded butterfly wings
	Color: Bright pink to purple
	Size: 1/2 to 1 inch long
Bloom Time	April to August
Form/Foliage	Mounding habit with gray-green leaves consisting of 9 to 19 smaller leaflets (pinnately compound) that are fuzzy (pubescent) on both the upper and lower sides and are arranged opposite each other along a main stem (petiole)
Habitat	Foothills, sagebrush communities, and oak brush foothills
Elevation Range	4,000 to 7,500 feet
Look-Alike	Northern Sweetvetch (*Hedysarum boreale*) p. 168

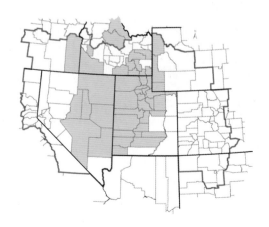

Long hairs on the round, pea-like seed
pods help deter insects from feeding.

Northern Sweetvetch
Hedysarum boreale

Plant Size	6 to 24 inches
Flower	Clusters of pea-like flowers alternate along a single stalk (raceme) and resemble large folded butterfly wings
	Color: Pink to magenta Size: 1/2 to 3/4 inch long
Bloom Time	June to July
Form/Foliage	Mounding habit with leaves (green above and gray beneath) consisting of 7 to 15 smaller leaflets (pinnately compound) that are arranged opposite each other along a main stem (petiole)
Habitat	Dry hillsides, sagebrush communities, and dry, open slopes
Elevation Range	4,100 to 9,100 feet
Look-Alike	Western Sweetvetch (*Hedysarum occidentale*)

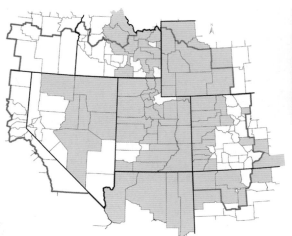

Look-Alike drooping flowers

Historically,
Americans
the roots, w
supposedly taste
like licorice.

Steer's Head
Dicentra uniflora

Plant Size	1 to 2 inches
Flower	Solitary flowers comprising 4 petals, with the upper 2 flaring upward and resembling horns
	Color: Pink to magenta Size: About 1/2 inch wide
Bloom Time	March to June
Form/Foliage	Small, dainty habit with sparse, deeply lobed (3 to 5) basal leaves
Habitat	Rocky slopes, gravelly hillsides, and sagebrush communities
Elevation Range	4,900 to 7,200 feet

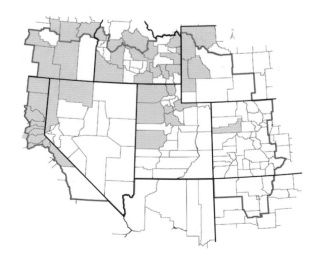

There is no chance to mistake this
unique bovine beauty, but you do need
to look low and close to see it.

Sticky Geranium
Geranium viscosissimum

Plant Size	16 to 30 inches
Flower	Solitary 5-petaled, distinctly veined flowers, borne atop branched stalks
	Color: Pink to magenta with dark veins Size: 1 to 1 1/2 inches wide
Bloom Time	May to September
Form/Foliage	Mounding habit with dark green, deeply dissected, palm-like (palmately compound) sticky leaves and stems
Habitat	Moist meadows, open woodlands, and sagebrush communities
Elevation Range	5,100 to 10,500 feet
Look-Alike	Pineywoods Geranium (*Geranium caespitosum*)

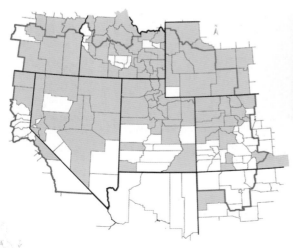

Look-Alike shorter, diminutive form with more leaf lobes that are rounded

The name *viscosissimum*
means "very sticky."

Nettleleaf Giant Hyssop
Agastache urticifolia

Plant Size	3 to 6 feet
Flower	Clusters of individual tubular flowers arranged in whorls (spike) that create an elongated, conical head that resembles a bottlebrush
	Color: Pink to lavender Size: 1 1/2 to 4 inches long (cluster)
Bloom Time	June to August
Form/Foliage	Upright habit with broad, triangular, serrated, aromatic leaves that are found opposite each other as they radiate up square stems
Habitat	Open woodlands, moist to dry meadows, and along stream banks
Elevation Range	5,400 to 9,800 feet

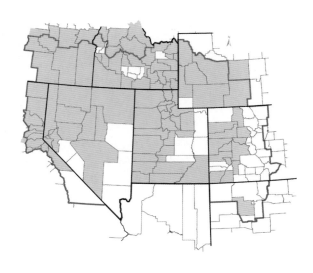

Also called Horsemint. A distinguishing
feature of all members of the mint
family is square stems.

Wild Bergamot
Monarda fistulosa

Plant Size	1 to 2 1/2 feet
Flower	Loose clusters (made up of individual flowers) are borne on terminal stalks and are cradled underneath by cupping leaves (bracts)
	Color: Pink Size: 1 to 3 inches wide (cluster)
Bloom Time	June to August
Form/Foliage	Upright habit with hairy (pubescent) oval (pointed tip) leaves that are found opposite each other on a square stem
Habitat	Moist meadows and shaded woodlands
Elevation Range	6,300 to 6,850 feet

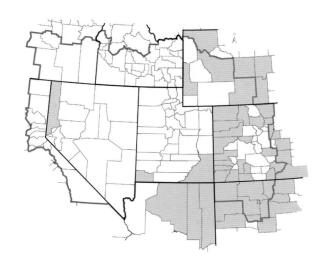

Also called Bee
Balm because it is
a favored plant of
bees, butterflies,
and hummingbirds.

Tapertip Onion
Allium acuminatum

Plant Size	6 to 12 inches
Flower	Umbrella-like clusters (umbels) consisting of 6-pointed flowers, borne atop a single stalk
	Color: Pink to white Size: 1/4 to 1/2 inch wide
Bloom Time	May to June
Form/Foliage	Narrow, upright habit with 2 to 4 grass-like basal leaves
Habitat	Dry hillsides and plains
Elevation Range	4,000 to 8,200 feet
Look-Alike	Nodding Onion (*Allium cernuum*)

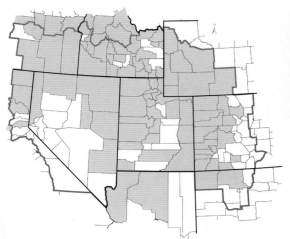

Look-Alike drooping flower head

Tapertip Onion can also be identified by its garlic-like aroma when the stem is crushed.

Mountain Wild Hollyhock
Iliamna rivularis

Plant Size	1 1/2 to 6 1/2 feet
Flower	Solitary flowers, consisting of 5 large silky petals, that radiate along a central stalk (raceme)
	Color: Pink to rose Size: 2 to 2 1/2 inches wide
Bloom Time	June to October
Form/Foliage	Upright habit with dark green, maple-like leaves (3 to 7 distinct lobes) that are 2 to 8 inches wide and alternate along the stems
Habitat	Stream banks, dry riverbeds, and open meadows
Elevation Range	5,400 to 9,400 feet

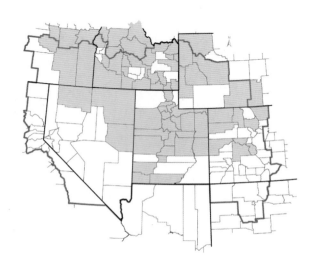

The seeds break apart like orange wedges when mature.

Oregon Checkermallow
Sidalcea oregana

Plant Size	16 to 50 inches
Flower	Loose clusters of 5-petaled silky flowers borne along a single stalk (spike)
	Color: Pink to rose Size: 1/2 to 1 3/4 inches wide
Bloom Time	May to August
Form/Foliage	Tall, upright habit with dark green, deeply divided, palm-like (palmately compound) leaves that alternate along the stems
Habitat	Moist meadows and stream banks
Elevation Range	4,000 to 8,900 feet

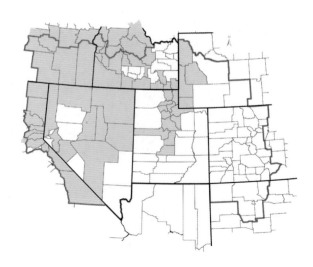

Oregon Checkermallow is a very attractive plant that resembles the old-fashioned garden variety hollyhocks (*Alcea rosea*).

Fireweed
Chamerion angustifolium

Plant Size   2 to 6 feet

Flower   Tubular, 4-petaled flowers that radiate along an erect stalk (raceme)

Color: Pink to magenta
Size: 1/2 to 5/8 inch long

Bloom Time   June to September

Form/Foliage   Upright habit with dark green, sword-like leaves that alternate along the red-tinged stems

Habitat   Open woodlands, moist meadows, and recently burned/disturbed sites

Elevation Range   5,000 to 11,200 feet

This plant is called "Fireweed" because
it is one of the first plants to populate
large areas after forest fires.

Longleaf Phlox
Phlox longifolia

Plant Size	2 to 6 inches
Flower	Loose clusters of tubular flowers with 5 petals that flare outward at right angles from the small throat
	Color: Pink to white Size: 1/ 2 inch wide
Bloom Time	April to July
Form/Foliage	Low-growing, airy habit with linear, woolly leaves found opposite each other along the stems
Habitat	Dry slopes and sagebrush communities
Elevation Range	4,500 to 10,400 feet
Look-Alike	Moss Campion (*Silene acaulis*) p. 162

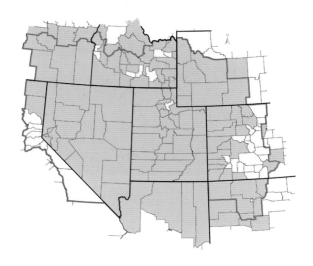

The word *phlox* in Greek means "fire," referring to the myriad of bright flowers that cover the plant.

Cushion Buckwheat
Eriogonum ovalifolium

Plant Size   3/4 to 2 inches

Flower   Blossoms arranged in a tight pompom-like
                        cluster (umbel), borne on a single stalk above
                        the foliage

                        Color: Pink to white
                        Size: 3/4 to 1 inch wide (cluster)

Bloom Time   June to August

Form/Foliage   Low-growing, mounding habit with gray-green,
                        oval leaves that form a tight basal rosette

Habitat   Dry, gravelly slopes, scree fields, and
                        sagebrush communities

Elevation Range   4,150 to 11,950 feet

The species name *ovalifolium* refers to
the oval shape of the leaves.

Darkthroat Shootingstar
Dodecatheon pulchellum

Plant Size	12 to 15 inches
Flower	1 to several solitary flowers, with petals that curve backward (reflex) and resemble a shooting comet, borne on naked stalks
	Color: Pink to violet, with white and yellow bands on the throat Size: 1/4 to 1/2 inch long
Bloom Time	May to July
Form/Foliage	Dainty habit with wide oval leaves that form a basal rosette
Habitat	Moist meadows and stream banks
Elevation Range	4,500 to 10,900 feet

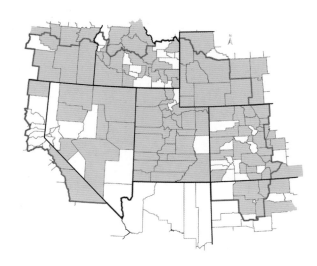

This plant's flowers are pollinated by the vibration of hovering bumblebees.

Parry's Primrose
Primula parryi

Plant Size	4 to 12 inches
Flower	Loose clusters of 5-petaled flowers, lobed at their tips, borne on branched stalks
	Color: Pink to violet, with yellow eye Size: 3/4 to 1 inch wide
Bloom Time	June to July
Form/Foliage	Loosely upright habit with smooth (glabrous), oval (pointed tip) leaves that are 8 to 10 inches long and form a tufted rosette at the base of the plant
Habitat	Rocky outcroppings and following receding snowfields
Elevation Range	8,000 to 12,900 feet

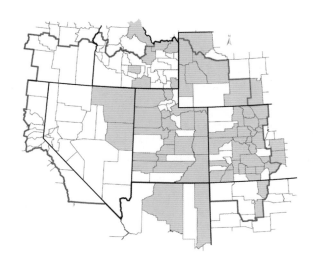

This plant produces flowers that smell like rotting meat (carrion) to attract pollinating flies.

Wildflowers of the Mountain West

Cliff Anemone
Anemone multifida

Plant Size	12 to 18 inches
Flower	Solitary flowers with 5 to 6 petals (sepals), borne atop a terminal stalk
	Color: Purple, red, and occasionally white Size: 3/4 to 1 inch wide
Bloom Time	April to August
Form/Foliage	Mounding habit with hairy (pubescent) leaves (grouped in 3s) arranged in a rosette at the base of the plant
Habitat	Open woodlands, rocky slopes, and meadows
Elevation Range	6,700 to 9,400 feet
Look-Alike	There are multiple colors and varieties found in the Mountain West

The name *Anemone* is Greek for "flower shaken by the wind," thus the often-used common name of Windflower.

Old Man's Whiskers
Geum triflorum

Plant Size	6 to 16 inches
Flower	Loose clusters (cymes) of 3 drooping, bulb-like flowers that become erect as the flowers mature, producing plume-like seed heads
	Color: Pink
	Size: 1/2 inch wide
Bloom Time	June to August
Form/Foliage	Loosely mounding habit with bright green, fern-like leaves found primarily at the base of the plant
Habitat	Moist meadows, stream banks, and aspen woodlands
Elevation Range	4,950 to 11,450 feet

The specific name *triflorum* is derived from the fact that each stalk gives rise to three unique flowers.

Wild Coralbells
Heuchera rubescens

Plant Size	4 to 14 inches
Flower	Clusters of tiny bell-shaped flowers that alternate along wiry, spike-like stalks (racemes)
	Color: Pink to red to white Size: Less than 1/4 inch
Bloom Time	April to September
Form/Foliage	Mounding habit with dark green, rounded leaves that have scalloped edges (consisting of 5 to 10 lobes) and form a rosette at the base of the plant
Habitat	Rocky outcroppings and shady sides of cliffs
Elevation Range	4,500 to 11,000 feet

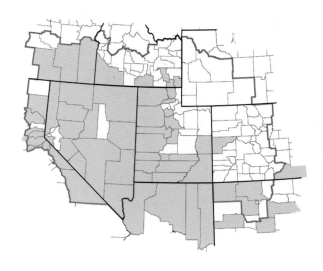

This plant was first collected on the summit of Stansbury Island (Utah) in 1850.

Lewis's Monkeyflower
Mimulus lewisii

Plant Size	10 inches to 3 feet
Flower	Unique face-like (bilaterally symmetrical), tubular flowers borne in pairs in the axils of the leaves
	Color: Pink to rose, with yellow throat Size: 3/4 to 1 inch wide
Bloom Time	July to September
Form/Foliage	Upright habit with dark green, oval (pointed tip), toothed leaves (with 3 to 7 parallel veins) found opposite each other on the stems
Habitat	Streams banks and moist, shady meadows
Elevation Range	4,600 to 11,400 feet
Look-Alike	Seep Monkeyflower (*Mimulus guttatus*) p. 124

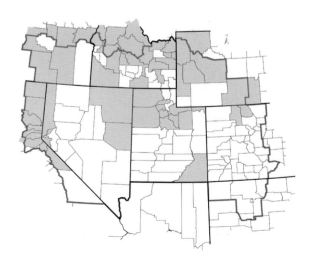

This plant was named after the famed western explorer Meriwether Lewis (Lewis and Clark Expedition).

Palmer's Beardtongue
Penstemon palmeri

Plant Size	1 1/2 to 4 1/2 feet
Flower	Several tiers consisting of 2 to 4 tubular flowers per tier (cyme), borne loosely along 1 side of a stalk (secund)
	Color: Pink to white, with violet stripes Size: 1 to 1 3/8 inches long
Bloom Time	May to July
Form/Foliage	Upright habit with blue-green, thick, smooth, oval (pointed tip) leaves that are opposite and adjoined to each other along the stem
Habitat	Along roadsides and in sagebrush communities, dry washes, and coniferous forests
Elevation Range	2,600 to 8,950 feet

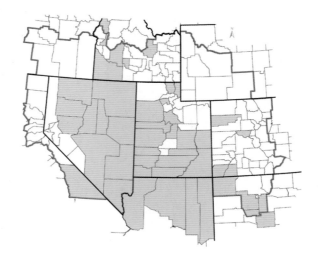

This beardtongue should be easily recognized
by the opposite leaves that appear as one
because they envelop the stem.

PURPLE FLOWERS

Western Monkshood (*Aconitum columbianum*), Utah
County, Utah, 2010, Elevation: 6,313 ft.

Aspen Fleabane
Erigeron speciosus

Plant Size	6 inches to 2 1/2 feet
Flower	Multiple clusters of daisy-like (composite) flowers with 60 to 150 narrow outer petals (ray flowers), borne on a single stalk
	Color: Purple to nearly white
	Size: 1 1/2 to 2 inches wide
Bloom Time	June to August
Form/Foliage	Upright, open habit with gray-green, oval (pointed tip), hairy (pubescent) leaves that alternate along the stem
Habitat	Open woodlands and foothills
Elevation Range	6,500 to 11,000 feet

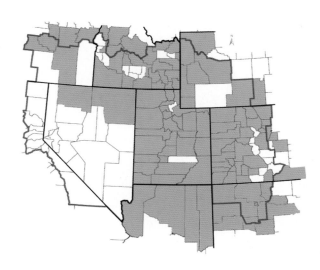

Asters have several overlapping rows of leaf-like structures (bracts) at the base of the flowers (involucre), much like shingles on a roof, whereas fleabane daisies have only one to two rows.

Roundleaf Harebell
Campanula rotundifolia

Plant Size	5 to 16 inches
Flower	Drooping (pendulous), bell-shaped flowers that alternate along the stalk
	Color: Purple to lavender Size: 1/2 to 1 inch long
Bloom Time	June to September
Form/Foliage	Upright habit with rounded basal leaves and linear leaves that alternate along erect stems
Habitat	Moist woodlands, grassy meadows, and roadsides
Elevation Range	5,300 to 12,000 feet

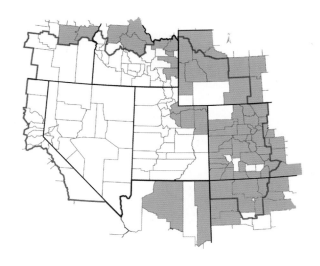

After seeing this little "belle" of a flower, you will not wonder that there are hundreds of its relatives in cultivated landscapes.

Silvery Lupine
Lupinus argenteus

Plant Size   6 inches to 2 1/2 feet

Flower   Individual pea-like flowers that are borne along a tall, vertical stalk (raceme) and develop into distinctive, hairy, pea-like seedpods

Color: Purple to lavender
Size: 1/4 to 1/2 inch long

Bloom Time   June to August

Form/Foliage   Upright, dense habit with gray-green, hairy (pubescent), palm-like leaves (palmately compound) that consist of 5 to 10 leaflets that are attached to a long stem (petiole)

Habitat   Open woodlands, dry foothills, and meadows

Elevation Range 5,000 to 10,900 feet

Though lupines are some of the most commonly encountered wildflowers in the Mountain West, they were unknown to science until Lewis and Clark collected a specimen in Montana in July of 1806.

Prairie Lupine
Lupinus lepidus

Plant Size	2 to 20 inches
Flower	Individual pea-like flowers that are borne along a vertical stalk (raceme), which can be immersed within the foliage, and develop into distinctive, hairy, pea-like seedpods
	Color: Purple to lavender Size: 1/8 to 3/8 inch long
Bloom Time	June to August
Form/Foliage	Low, mounding habit with gray-green, hairy (pubescent), palm-like leaves (palmately compound) that consist of 5 to 10 leaflets that are attached to a short stem (petiole)
Habitat	Gravelly slopes and disturbed sites
Elevation Range	6,200 to 11,150 feet
Look-Alike	*Lupinus lepidus* var. *lobbii*

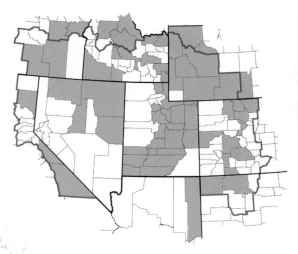

Look-Alike flowers extend
beyond the leaves

There are ten varieties of *Lupinus lepidus* in the western United States, with variety *utahensis* being the most widespread.

Ballhead Waterleaf
Hydrophyllum capitatum

Plant Size	4 to 16 inches
Flower	Multiple tubular flowers arranged in round clusters (cymes) of puffball-like heads that are found below the foliage
	Color: Purple to lavender Size: 1 to 2 inches wide (cluster)
Bloom Time	April to July
Form/Foliage	Mounding habit with bright green, succulent, hairy (pubescent) leaves that are deeply lobed with 7 to 10 leaflets
Habitat	Open woodlands and moist slopes
Elevation Range	4,600 to 9,000 feet

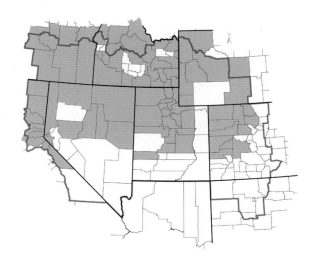

The tender shoots and roots were used by Native Americans and early western settlers as a potherb.

Silverleaf Phacelia
Phacelia hastata

Plant Size	6 inches to 3 feet
Flower	Clusters of individual flowers that are borne along a tightly coiled stalk and become more evident and upright as they open with maturity
	Color: Purple to white Size: 1/4 inch long
Bloom Time	May to July
Form/Foliage	Upright habit with gray-green, spear-like, hairy (pubescent) leaves that alternate along the stems
Habitat	Open woodlands, gravelly hillsides, and dry, open slopes
Elevation Range	4,400 to 11,400 feet

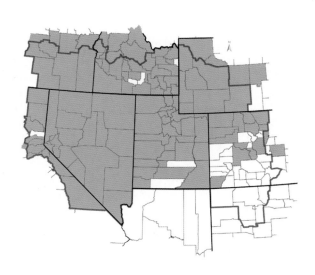

The species name *hastata* refers to the "arrowhead" shape of the leaves.

Purple Fringe
Phacelia sericea

Plant Size	4 to 20 inches
Flower	Multiple tiny, tubular flowers, with exaggerated yellow, whisker-like flower parts (anthers) arranged in a bottlebrush-like cluster (thyrse)
	Color: Purple Size: 4 to 6 inches long (cluster)
Bloom Time	June to August
Form/Foliage	Upright habit with deeply lobed, hairy (pubescent) leaves that alternate along the stem, with most found primarily at the base of the plant
Habitat	Open woodlands and rocky slopes
Elevation Range	7,000 to 11,200 feet

The species name *sericea* means "silky," which refers to the fine hairs on the stems and leaves that give the plant a somewhat silky-smooth appearance...try squinting your eyes.

Western Iris
Iris missouriensis

Plant Size   1 to 2 feet

Flower   1 to 3 flowers, consisting of 3 downturning petals (falls) and 3 upright petals (tepals) that are streaked with yellow and white markings, borne atop a single stalk

Color: Purple to light lavender
Size: 2 to 3 inches long

Bloom Time   May to July

Form/Foliage   Upright habit with long, linear, grass-like basal leaves

Habitat   Moist meadows, stream banks, marshes, and aspen woodlands

Elevation Range   5,400 to 9,450 feet

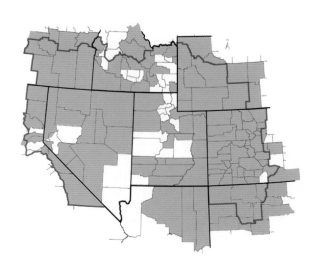

The downturning petals (falls) serve as landing pads for pollinators.

Douglas's Grasswidow
Olsynium douglasii

Plant Size 1 to 1 1/2 feet

Flower 1 to 3 star-shaped flowers consisting of 6 petals (tepals), borne atop a single stalk

Color: Blue to purple, with a yellow center
Size: 1/2 to 3/4 inch wide

Bloom Time March to June

Form/Foliage Upright habit with narrow, grass-like leaves

Habitat Grassy meadows, foothills, and gravelly slopes

Elevation Range 5,000 to 7,000 feet

Also called
Sisyrinchium, which
in Greek means
"swine snout."
Named after plants
whose roots were
favored by rooting
pigs.

Pale Monardella
Monardella odoratissima

Plant Size	3 to 18 inches
Flower	Globe-like clusters (made up of individual flowers) that are borne on terminal stalks and are cradled underneath by colorful leaves (bracts) Color: Purple to light pink Size: 1/2 to 1 inch wide (cluster)
Bloom Time	June to September
Form/Foliage	Mounding habit with smooth, oval (pointed tip) leaves that are found opposite each other on a square stem
Habitat	Dry hillsides, slopes, and open, gravelly sites
Elevation Range	5,900 to 11,000 feet

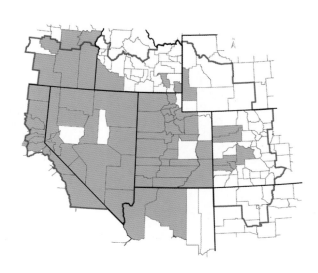

Also called Mountain Pennyroyal, this flower has a distinct mint odor when disturbed.

Western Monkshood
Aconitum columbianum

Plant Size	2 to 4 feet
Flower	Unique hood-like flowers are created by a petal-like sepal (the hood) that sits above and envelops 2 smaller petals. These flowers alternate along a tall central stalk.
	Color: Deep purple Size: 3/4 to 2 inches long
Bloom Time	June to August
Form/Foliage	Upright habit with dark green, deeply dissected, palm-like (palmately compound) leaves that alternate along the stems and clasp directly onto the stems (sessile)
Habitat	Moist meadows, seeps or springs, and along stream banks
Elevation Range	5,700 to 11,000 feet
Look-Alike	Western Larkspur (*Delphinium ×occidentale*) p. 262

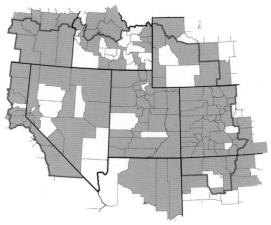

Pollinated by bumblebees, which are the only insects strong enough to push the petals apart.

Hairy Clematis
Clematis hirsutissima

Plant Size   1 to 2 feet

Flower   Solitary flowers, each with 4 petals (sepals) fused together to form a hairy bell, borne on a single stalk

                      Color: Purple
                      Size: 1 to 1 1/2 inches long

Bloom Time   June to July

Form/Foliage   Upright, erect habit with hairy (pubescent), deeply dissected (pinnately compound) leaves that are found opposite each other on the stems

Habitat   Moist meadows, sagebrush communities, and spruce/fir woodlands

Elevation Range   6,700 to 10,900 feet

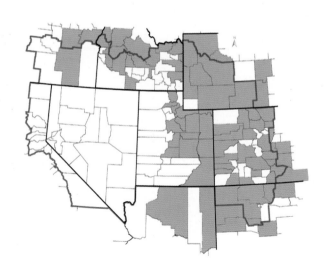

This plant, unlike most other clematis species, is not a vine.

Spent flowers and seed heads.

Elephanthead Lousewort
Pedicularis groenlandica

Plant Size   6 inches to 2 feet

Flower   Unique individual flowers that are shaped like
                        an elephant's head (complete with ears and
                        an upturned trunk) and are arranged in tight
                        clusters along a single stalk (spike)

                        Color: Purple to rose
                        Size: 1 1/2 to 6 inches long (cluster)

Bloom Time   July to August

Form/Foliage   Dainty habit with dark green, feathery, fern-
                        like (pinnately compound) leaves that alternate
                        along the stems, with most found primarily at
                        the base of the plant

Habitat   Moist meadows, bogs, and stream banks

Elevation Range   8,400 to 11,900 feet

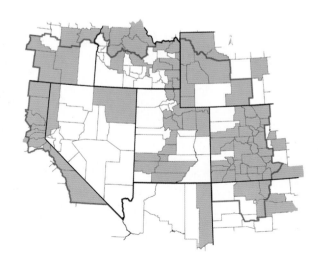

Although rare, there are reports of white variations of this plant.

Broadleaf Beardtongue
Penstemon platyphyllus

Plant Size	1 to 2 feet
Flower	4 to 8 tiered whorls (cymes) consisting of 1 to 2 tubular flowers, borne along an upright stalk
	Color: Pale purple to lavender Size: 3/4 to 1 inch long
Bloom Time	June to July
Form/Foliage	Shrubby habit of ascending branches with smooth, sword-like leaves found opposite each other on the stems
Habitat	Rocky slopes and dry hillsides
Elevation Range	4,900 to 8,900 feet
Look-Alike	Other beardtongues (*Penstemon* spp.)

This beardtongue is found only between Ogden Canyon and American Fork Canyon in the Wasatch Range of northern Utah.

Rocky Mountain Beardtongue
Penstemon strictus

Plant Size	1 to 2 1/2 feet
Flower	4 to 10 tiers consisting of 1 to 2 tubular flowers per tier (cyme), borne along 1 side of a stalk (secund)
	Color: Purple to deep blue
	Size: 1 to 1 1/4 inches long
Bloom Time	June to August
Form/Foliage	Upright habit with smooth, narrow leaves that clasp directly onto the stem (sessile) and are found opposite each other
Habitat	Open meadows, sagebrush communities, and aspen/conifer forests
Elevation Range	6,750 to 10,800 feet

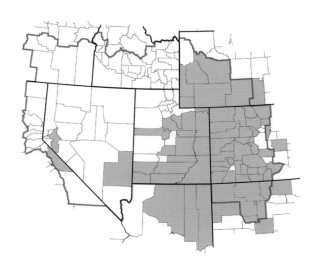

Beardtongues are the third largest group of wildflowers in North America.

Wildflowers of the Mountain West

Whipple's Beardtongue
Penstemon whippleanus

Plant Size   6 to 24 inches

Flower   2 to 5 closely spaced tiers or whorls (cymes)
                        consisting of 2 to 4, tubular flowers that are
                        borne on 1 side (secund) of an upright stalk

                        Color: Deep purple to white, with a white-
                        striped beard in the throat
                        Size: 1/2 inch long

Bloom Time   July to August

Form/Foliage   Low-growing, multistemmed habit with sword-
                        like leaves that are found opposite each other
                        on the stem and clustered at the base of the
                        plant

Habitat   Open woodlands and dry slopes

Elevation Range   5,800 to 11,600 feet

Look-Alike   Other beardtongues (*Penstemon* spp.)

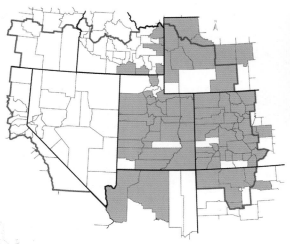

Look-Alike a yellow
variation

Possibly the deepest purple of all the wildflowers; the lower petal has extremely long, white hairs near the throat.

Hooked Violet
Viola adunca

Plant Size	6 to 10 inches
Flower	Solitary 5-petaled flowers (consisting of 3 basal petals and 2 upward-pointing petals) borne in the leaf axils
	Color: Purple to lavender Size: 1 to 1 1/2 inches wide
Bloom Time	March to August
Form/Foliage	Small, mounding to spreading habit with heart-shaped basal leaves
Habitat	Shady woodlands and grassy meadows
Elevation Range	4,800 to 9,900 feet

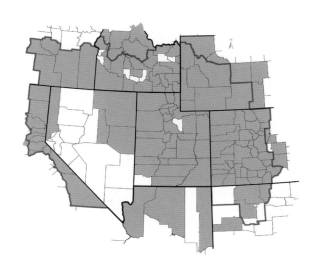

The purple venation on the lower flower petal serves as guidelines or a "runway" to direct incoming pollinators.

BLUE FLOWERS

**Prairie Flax (*Linum lewisii*), Cache County, Utah, 2010
Elevation: 8,273 ft.**

Alpine Forget-me-not
Eritrichium nanum

Plant Size	2 to 4 inches
Flower	Tight cluster of 5-petaled flowers borne atop a stalk
	Color: Blue, often with a yellow center
	Size: 1/4 inch wide
Bloom Time	July to August
Form/Foliage	Low-growing habit with short, oval, woolly leaves that are primarily basal, with a few alternating along compressed stems
Habitat	Open, rocky slopes and alpine tundra
Elevation Range	9,900 to 13,000 feet

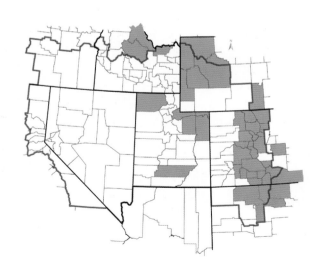

The short hairs that cover the entire plant help regulate temperature and moisture.

Meadow Forget-me-not
Hackelia micrantha

Plant Size	1 to 3 1/2 feet
Flower	Clusters of loosely arranged 4- to 5-petaled flowers (each petal has a slightly raised fold at the base) borne on single stalks
	Color: Light blue to lavender
	Size: 1/4 to 1/2 inch wide
Bloom Time	June to August
Form/Foliage	Airy habit with sword-like, hairy (pubescent) leaves that alternate along wiry stems
Habitat	Open woodland meadows and stream banks
Elevation Range	6,900 to 9,900 feet
Look-Alike	Stickseed (*Hackelia patens*)

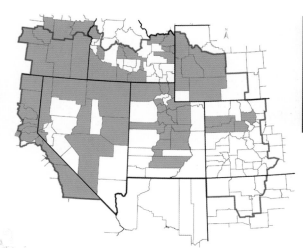

Look-Alike white, at lower elevations

Also called Stickseed because of the tiny barbs on the seeds that stick to animals and clothing.

Tall Mountain Bluebell
Mertensia ciliata

Plant Size	2 to 4 feet
Flower	Multiple clusters of tubular, bell-shaped flowers (5 fused petals) that are borne along a single stalk (cyme) and have a drooping (pendulous) habit
	Color: Blue to purple to pink Size: 1/2 to 3/4 inch long
Bloom Time	June to August
Form/Foliage	Upright habit with smooth, blue-green, oval (pointed tip) leaves that have fine hairs on the margins and alternate along the stem
Habitat	Open woodlands, moist meadows, and stream banks
Elevation Range	6,500 to 12,400 feet

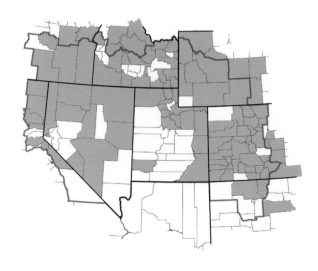

This plant is a favorite browse (food) for elk and deer.

Leafy Bluebell
Mertensia oblongifolia

Plant Size	4 to 16 inches tall
Flower	Multiple clusters of tubular, bell-shaped flowers (5 fused petals) that are borne along a single stalk (cyme) and have a drooping (pendulous) habit
	Color: Blue Size: 3/8 to 3/4 inch long
Bloom Time	May to July
Form/Foliage	Low, tufted habit with smooth, blue-green, spear-like leaves that arise at the base of the plant
Habitat	Open slopes, sagebrush communities, and dry meadows
Elevation Range	4,650 to 9,950 feet
Look-Alike	Lanceleaf Bluebell (*Mertensia lanceolata*)

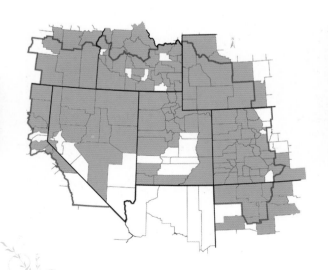

Look-Alike at higher elevations

Bluebells are pollinated by bees, flies, and butterflies.

Mountain Gentian
Gentiana calycosa

Plant Size   3 to 8 inches

Flower   Solitary broadly tubular flowers, each with
                    flared and flattened petal tips (which give them
                    the appearance of an inverted bell), borne atop
                    single stalks

                    Color: Blue
                    Size: 1 to 1 1/2 inches long

Bloom Time   July to September

Form/Foliage   Mounding habit with glossy green, rounded
                    leaves that are found opposite each other on
                    the stems

Habitat   Moist meadows and stream banks

Elevation Range 7,200 to 10,500 feet

This plant is also called Pleated Gentian, referring to the folds or "pleated" connections where the petals are fused together.

Prairie Flax
Linum lewisii

Plant Size	6 to 30 inches
Flower	Loose clusters (racemes) of 5-petaled flowers borne terminally on a stalk
	Color: Blue Size: 1/2 to 1 inch wide
Bloom Time	June to September
Form/Foliage	Airy habit with small, narrow, oval leaves that alternate along the stem
Habitat	Rocky slopes, gravelly hillsides, and disturbed sites
Elevation Range	3,900 to 10,300 feet
Look-Alike	King's Flax (*Linum kingii*)

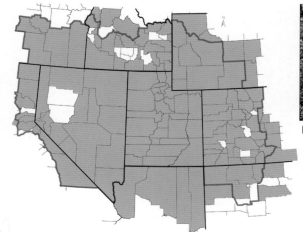

Look-Alike yellow-flowered, low-growing species

This plant was named after the famed
western explorer Meriwether Lewis
(Lewis and Clark Expedition).

Showy Jacob's Ladder
Polemonium pulcherrimum

Plant Size   2 to 18 inches

Flower   Clusters of 5-petaled flowers, each with 5 white pollen-laden structures (anthers), borne atop branched stalks

Color: Blue
Size: 1/3 to 2/3 inch wide

Bloom Time   June to August

Form/Foliage   Loose, mounding habit with bright green, dissected leaves made up of smaller leaflets (pinnately compound) that resemble rungs on a ladder

Habitat   Shady woodlands and stream banks

Elevation Range 4,400 to 9,800 feet

Look-Alike   Western Polemonium
(*Polemonium occidentale*)

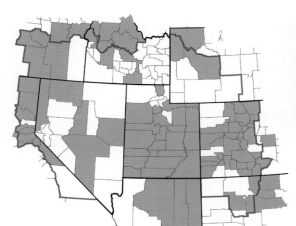

Look-Alike a taller plant with yellow anthers

The species name *pulcherrimum* means "most beautiful."

Sky Pilot
Polemonium viscosum

Plant Size	4 to 12 inches
Flower	Clusters of 5-petaled flowers, each with 5 yellow pollen-laden structures (anthers), borne atop branched stalks
	Color: Blue
	Size: 1/2 to 1 inch wide
Bloom Time	June to August
Form/Foliage	Tight, mounding habit with numerous tiny, round, hairy (pubescent) leaves that are crowded (radially) along the stems
Habitat	Rocky slopes and open hillsides at high elevations
Elevation Range	9,000 to 12,900 feet

This plant is also called Sticky Polemonium, referring to the "tacky" stems and leaves.

Colorado Columbine
Aquilegia caerulea

Plant Size   18 to 30 inches

Flower   Intricate drooping (pendulous) flowers
                     consisting of 5 petals (spurs) and 5 alternating
                     sepals, which are flattened or flared, borne
                     atop a single stalk

                     Color: White to light blue
                     Size: 3 to 4 inches wide

Bloom Time   July to August

Form/Foliage   Airy habit with bright green, thin leaves that are
                     distinctly 3-lobed (trifoliate) and found primarily
                     at the base of the plant

Habitat   Open woodlands, high, rocky slopes, and
                     stream banks

Elevation Range 5,400 to 11,900 feet

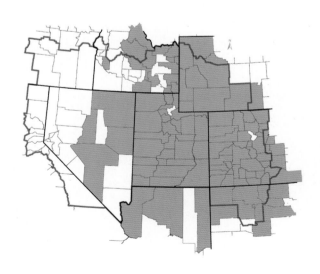

This is the state flower of Colorado.

Lowland Larkspur
Delphinium nuttallianum

Plant Size	4 to 16 inches
Flower	Individual deep-throated, tubular flowers (resembling a wizard's hat) that alternate along a main stalk (raceme) Color: Blue Size: 3/8 to 3/4 inch long
Bloom Time	June to August
Form/Foliage	Few leaves that are deeply dissected and palm-like (palmately compound) and alternate along short stems
Habitat	Dry meadows, sagebrush communities, and open woodlands
Elevation Range	3,950 to 10,700 feet

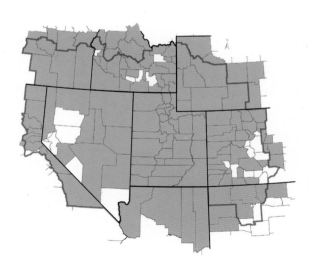

Delphinium in Greek means "dolphin flower" and refers to the shape of the early flower bud.

Western Larkspur
Delphinium ×occidentale

Plant Size	4 to 6 feet
Flower	Individual deep-throated, tubular flowers (resembling a wizard's hat) that alternate along a main stalk (raceme)
	Color: Blue to lavender
	Size: About 1 inch long
Bloom Time	June to August
Form/Foliage	Upright habit with dark green, deeply dissected, palm-like (palmately compound) leaves that alternate along the stems and clasp directly onto the stems (sessile)
Habitat	Open meadows, bogs, and stream banks
Elevation Range	5,100 to 10,900 feet
Look-A-Like	Western Monkshood (*Aconitum columbianum*) p. 226

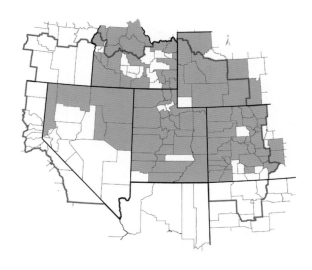

D. occidentale is a hybrid resulting from the convergence of *D. glaucum* from the Northwest and *D. barbeyi* from the East and is more prominent than either of its parent species.

Wasatch Beardtongue
Penstemon cyananthus

Plant Size	2 to 3 feet
Flower	2 to 5 denselytiered whorls (cymes) consisting of 3 to 7 tubular flowers, borne along an upright stalk
	Color: Blue to purple Size: 3/4 to 1 1/4 inches long
Bloom Time	May to July
Form/Foliage	Upright habit with smooth, oval-shaped leaves that clasp directly onto the stem (sessile) and are found opposite each other
Habitat	Gravelly hillsides, dry slopes, and disturbed sites
Elevation Range	5,000 to 10,800 feet
Look-Alike	Low Beardtongue (*Penstemon humilis*)

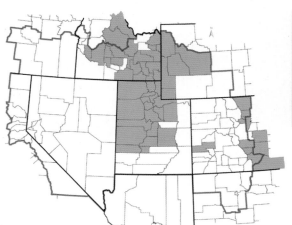

Look-Alike shorter, darker blue species

There are over 100 individual species of beardtongue documented in the Mountain West.

GREEN FLOWERS

**Green Gentian (*Frasera speciosa*), Utah County, Utah, 2010
Elevation: 7,357 ft.**

Green Gentian
Frasera speciosa

Plant Size	3 to 6 feet
Flower	Solitary 4-petaled flowers, originating in the leaf axil, with deep nectar reservoirs (glands) and unique purple to brown freckles that help attract pollinating insects
	Color: Green to white Size: 3/4 to 1 1/2 inches wide
Bloom Time	June to August
Form/Foliage	Upright form with narrow, spear-like leaves (10 to 20 inches long) that are arranged in circular whorls and become less prominent as they ascend up the tower-like stalk
Habitat	Open woodlands, moist meadows, and sagebrush communities
Elevation Range	6,500 to 11,300 feet
Look-Alike	False Hellebore (*Veratrum californicum*)

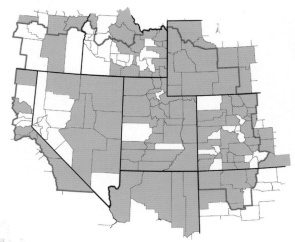

Look-Alike alternate leaves, flowers in tight clusters

This plant can live
multiple years but
dies once it blooms.

BROWN FLOWERS

Western Coneflower (*Rudbeckia occidentalis*), Franklin County, Idaho, 2009, Elevation: 6,541 ft.

Western Coneflower
Rudbeckia occidentalis

Plant Size	3 to 6 feet
Flower	Distinctive dome-shaped flowers, lacking petals (rayless composites), borne singly on a long, thick stalk
	Color: Brown Size: 1 to 2 inches long
Bloom Time	July to August
Form/Foliage	Upright habit with rough, sandpaper-like, oval (pointed tip) leaves (3 to 10 inches long) that alternate along a stout stem
Habitat	Open meadows, woodlands, and stream banks
Elevation Range	6,200 to 9,900 feet

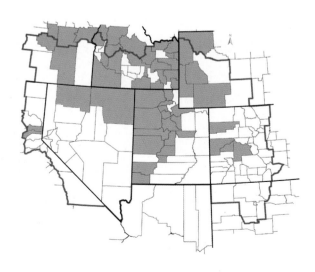

As deceiving as it may seem, this unattractive flower is a favorite of bees because of its high nectar and pollen content.

Leopard Lily
Fritillaria atropurpurea

Plant Size	4 to 24 inches
Flower	Solitary nodding (pendulous), 6-petaled (tepaled) flowers, borne terminally on a single stalk
	Color: Brown to purple, with yellow spots Size: 1 to 1 1/2 inches wide
Bloom Time	April to July
Form/Foliage	Dainty habit with bright green, narrow, grass-like leaves that alternate along the stem
Habitat	Rocky slopes and open woodlands
Elevation Range	5,000 to 10,500 feet

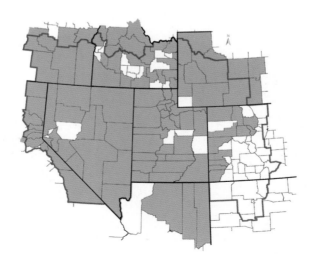

The brown coloration makes this flower difficult to see at first glance, which makes it even more exciting when you find it.

Hare Figwort
Scrophularia lanceolata

Plant Size   30 to 60 inches

Flower   Loose clusters of unique bulb-like, hollow
                        flowers with 2 erect, flattened petals that give
                        the appearance of rabbit ears

                        Color: Brown to yellow-green
                        Size: 1/3 to 1/2 inch long

Bloom Time   June to July

Form/Foliage   Upright habit with oval (pointed tip), serrated
                        leaves that are arranged opposite each other on
                        the stems

Habitat   Moist meadows, woodlands, and stream banks

Elevation Range   3,500 to 9,200 feet

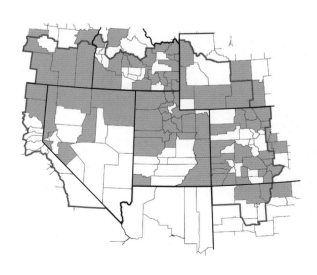

The uniquely fused petals help the flower look like a rabbit's face, making this a really fun plant for children to find.

Common Botanical Terms

Alternate:
a leaf arrangement in which the leaves arise from the stem in an alternating fashion from one another

Annual:
a plant that completes its life cycle (germinates, flowers, and dies) in a single growing season

Anther:
pollen-bearing portion of the stamen

Axil:
the point where the petiole of a leaf attaches to the stem

Basal:
leaves arranged or located at the base of a plant

Biennial:
a plant that germinates and grows vegetatively the first year, then flowers, sets seed, and dies the second year

Bract:
a modified leaf normally found below a flower

Calyx:
the collective sepals of a flower

Common name:
a regionally accepted name given to plants based on culture and use

Composite:
a daisy-like inflorescence composed of ray flowers (appear like petals) and disc flowers (found in the center). Both are small and packed together on a platform called a receptacle.

Compound:
a leaf composed of individual leaflets

Corm:
a short, compressed, underground stem

Corolla:
the collective petals of a flower

Corymb:
individual flower stalks growing from different points to approximately the same height, resulting in a flat-topped cluster. The outermost flowers open first.

Cyme:	branched flower stalks growing from the same point to approximately the same height, resulting in a flat-topped cluster. Central or inner flowers open first.
Disc flowers:	small, tubular-shaped flowers found in the center of a composite inflorescence
Falls:	the three downturning petals of an iris flower
Family:	a broad taxonomic ranking of plants of common lineage
Filament:	the long supportive structure of a stamen that bears the anther
Genus:	a taxonomic ranking of related species below the level of family (the first part of the scientific name of a species)
Glabrous:	lacking hair or pubescence
Inflorescence:	the collective arrangement of the flowers on a plant
Involucre:	the bracts surrounding and below the flower
Opposite:	a leaf arrangement in which the leaves arise from the stem in an opposing fashion from one another
Ovary:	the enlarged rounded portion of the pistil that contains the ovules (where the seeds will develop)
Palmate:	a type of compound leaf where individual leaflets emerge from a central point and appear like fingers radiating from a palm
Panicle:	a flower cluster composed of numerous branched cymes, corymbs, racemes, spikes, thyrses, or umbels
Pedicel:	a stalk that holds a single flower or each flower in an inflorescence
Pendulous:	drooping or hanging downward
Perennial:	a plant that lives for multiple seasons from the same growing point or root system

Petal:	the colorful, often flattened part of the flower, which attracts pollinators
Petiole:	the portion of the leaf where the blade attaches to the stem
Pinnate:	a type of compound leaf where individual leaflets emerge from a central axis and appear ladder-like
Pistil:	the female reproductive parts of a flower, consisting of the stigma, style, and ovary
Pubescent:	covered in fine hairs
Raceme:	multiple flowers, each with a stalk, or pedicel, attached to a central, indeterminate axis
Ray flower:	petal-shaped flowers radiating out from the center of a composite inflorescence
Reflexed:	a state of being bent or turned backward; also referred to as recurved
Rosette:	a consolidation of leaves at the base of a plant
Scientific name:	an internationally accepted Latin-based name given to plants
Scree:	a collective deposit of broken rock fragments associated with mountain slopes and cliffs
Secund:	flowers borne exclusively on one side of a stalk
Sepal:	a flower part that encloses and protects the bud
Sessile:	a leaf or flower that lacks a petiole or pedicel and connects directly to the stem or stalk
Solitary:	an individual flower
Species:	a group of individuals in a plant community that share physical characteristics and interbreed freely (the complete taxonomic ranking of a plant—genus and specific epithet collectively)
Specific epithet:	a taxonomic ranking of specific relation below the level of genus (the second part of a botanical name)

Common Botanical Terms

Spike:	multiple flowers that lack a stalk (also referred to as sessile) and attach to a central, indeterminate axis
Spur:	a sepal or petal that develops into an elongated hollow tube and extends behind the flower
Stalk:	the stem portion of the plant that bears the flowers, also known as a peduncle
Stamen:	the male reproductive part of the flower, consisting of the anther and filament
Stigma:	pollen-receiving end of the pistil
Style:	a hollow appendage that bears the stigma and contains the pollen tube
Succulent:	specialized leaves that are thick and contain water-filled cells
Tepal:	a petal or sepal of a flower in which there is no individual distinction between them
Thyrse:	a dense cluster of individual flowers arranged in a panicle-like fashion
Trifoliate:	a leaf consisting of three individual leaflets emerging from a central point
Tuber:	storage tissue, technically a stem, found underground on some plants
Umbel:	individual flower stalks growing from the same point to approximately the same height, resulting in a flat-topped cluster
Whorled:	a leaf arrangement in which multiple leaves arise from the stem from a single radial point

Bibliography

Albee, Beverly J., Leila M. Shultz, and Sherel Goodrich. *Atlas of the Vascular Plants of Utah*. Salt Lake City: Utah Museum of Natural History, 1988.

Andersen, Bernice A., and Arthur H. Holmgren. *Mountain Plants of Northeastern Utah*. Logan: Utah State University Extension Services Circular 319, 1970.

Anderson, Steve, Lisa Eriksen, Cheri Howell, Lois Ports, and Lisa Seymour. *Ruby Mountain Flora: A Guide to Common Plants of the Ruby Mountains and East Humboldt Range*. Elko, NV: Humboldt National Forest Interpretive Association, 1998.

Atlas of Colorado Vascular Plants. (n.d.). Retrieved from http://cumuseum. colorado.edu/Research/Botany/Databases/search.php.

Atlas of the Vascular Plants of Wyoming. (n.d.). Retrieved from http://www.sbs. utexas.edu/tchumley/wyomap/atlas.htm.

Axelrod, Daniel I., and Peter H. Raven. "Origins of the Cordilleran Flora." *Journal of Biogeography* 12:1 (Jan. 1985): 21–47. JSTOR. Utah State University Library, Logan: 9 Nov 2007. http://links.jstor.org/.

Barneby, Rupert C., Arthur Cronquist, Arthur H. Homgren, Noel H. Holmgren, James L. Reveal, and Patricia K. Holmgren. *Intermountain Flora: Vascular Plants of the Intermountain West, U. S. A*. Vol. 3, Part B. New York: New York Botanical Garden Press, 1989.

Bienz, D. R. *The Why and How of Home Horticulture*. New York: W. H. Freeman, 1980.

Billings, W. D. "Adaptations and Origins of Alpine Plants." *Arctic and Alpine Research* 6:2 (1974): 129–142. JSTOR. Utah State University Library, Logan: 14 Mar 2008. http://links.jstor.org/.

———. "Alpine Phytogeography Across the Great Basin." *Great Basin Naturalist Memoirs* 2 (1978): 105–117. Biodiversity Heritage Library. 21 Jun 2011. http://www.biodiversitylibrary.org/.

Bradley, W. Glen. "A Geographical Analysis of the Flora of Clark County, Nevada." *Journal of the Arizona Academy of Science* 4:3 (1967): 151–162. JSTOR. Utah State University Library, Logan: 9 Mar 2012. http://links.jstor.org/.

Brandenburg, David M. *Field Guide to Wildflowers of North America*. New York: Sterling Publishers, 2010.

Cain, Stanley A. *Foundations of Plant Geography*. New York: Harper and Brothers, 1944.

Chabot, Brian F., and W. D. Billings. "Origin and Ecology of the Sierran Alpine Flora and Vegetation." *Ecological Monographs* 42:2 (1972): 163–199. JSTOR. Utah State University Library, Logan: 18 Mar 2012. http://links.jstor.org/.

Charlet, David Alan. "Relationships of the Great Basin Alpine Flora: A Quantitative Analysis." Thesis. University of Nevada, Reno. 1991.

Clark, Stephen L. "A Manual of the Vascular Flora of Weber, Davis, and Morgan Counties:" Diss. Brigham Young University. 1980.

Consortium of California Herbaria. (n.d.). Retrieved from http://ucjeps.berkeley.edu/consortium/.

Consortium of Pacific Northwest Herbaria. (n.d.). Retrieved from http://www.pnwherbaria.org/resources.php#checklists.

Cronquist, Arthur. "The Biota of the Intermountain Region in Geohistorical Context." *Great Basin Naturalist Memoirs* 2 (1978): 3–15. Biodiversity Heritage Library. 9 Oct 2011. www.biodiversitylibrary.org.

Cronquist, Arthur, Arthur H. Holmgren, Noel H. Holmgren, and James L. Reveal. *Intermountain Flora: Vascular Plants of the Intermountain West, U. S. A.* Vol. 1. New York: Hafner, 1986.

Cronquist, Arthur, Arthur H. Holmgren, Noel H. Holmgren, James L. Reveal, and Patricia K. Holmgren. *Intermountain Flora: Vascular Plants of the Intermountain West, U. S. A.* Vol. 4. New York: New York Botanical Garden Press, 1984.

———. *Intermountain Flora: Vascular Plants of the Intermountain West, U. S. A.* Vol. 5. New York: New York Botanical Garden Press, 1994.

Cronquist, Arthur, Noel H. Holmgren, and Patricia K. Holmgren. *Intermountain Flora: Vascular Plants of the Intermountain West, U. S. A.* Vol. 3, Part A. New York: New York Botanical Garden Press, 1997.

Daubenmire, Rexford. *Plant Geography with Special Reference to North America*. New York: Academic Press, 1978.

Digital Atlas of the Vascular Plants of Utah. (n.d.). Retrieved from http://earth.gis.usu.edu/plants/.

Dodson, Carolyn, and William W. Dunmire. *Mountain Wildflowers of the Southern Rockies*. Albuquerque: University of New Mexico Press, 2007.

Dorn, Robert D. *Vascular Plants of Wyoming*. 2nd ed. Cheyenne, WY: Mountain West Publishing, 1992.

Earle, A. Scott, Jane Lundin, and James L. Reveal. *Idaho Mountain Wildflowers: A Photographic Compendium*. Boise, ID: Larkspur Books, 2008.

Ellison, Lincoln. "Subalpine Vegetation of the Wasatch Plateau, Utah." *Ecological Monographs* 24:2 (Apr. 1954): 89–184. JSTOR. Utah State University Library, Logan: 7 Feb 2008. http://links.jstor.org/.

Flora of North America Editorial Committee, eds. *Flora of North America North of Mexico*. 16+ vols. New York and Oxford: Oxford University Press, 1993+.

Hadley, Keith S. "Vascular Alpine Plant Distributions within the Central and Southern Rocky Mountains." *Arctic and Alpine Research* 19:3 (Aug. 1987): 242–251. JSTOR. Utah State University Library, Logan: 3 Jul 2011. http://links.jstor.org/.

Harper, K. T., D. Carl Freeman, W. Kent Ostler, and Lionel C. Klikoff. "The Flora of Great Basin Mountain Ranges: Diversity, Sources, and Dispersal Ecology." *Great Basin Naturalist Memoirs* 2 (1978): 81–103. Biodiversity Heritage Library. 8 Mar 2011. http://www.biodiversitylibrary.org/.

Harrington, H. D. *Manual of the Plants of Colorado*. Denver: Sage Books, 1964.

Harris, James G., and Melinda Woolf Harris. *Plant Identification Terminology: An Illustrated Glossary*. Spring Lake, UT: Spring Lake Publishing, 2001.

Hartman, Ronald L., and B. E. Nelson. *A Checklist of the Vascular Plants of Colorado*. Rocky Mountain Herbarium. Mar. 2001. Laramie: University of Wyoming. http://www.rmh.uwyo.edu.

Hayward, C. Lynn. "Biotic Communities of the Wasatch Chaparral, Utah." *Ecological Monographs* 18:4 (Oct. 1948): 473–506. JSTOR. Utah State University Library, Logan. 12 Dec 2010. http://links.jstor.org/.

Holmgren, Noel H., Patricia K. Holmgren, and Arthur Cronquist. *Intermountain Flora: Vascular Plants of the Intermountain West, U. S. A.* Vol. 2, Part B. New York: New York Botanical Garden Press, 2005.

Hooker, Joseph D. "On Geographical Distribution." *Proceedings of the Royal Geographical Society* 3 (1881): 594–608. Google Books. 17 Mar 2012. http://books.google.com/.

Irving, Washington. *The Adventures of Captain Bonneville*. New York: John B. Alden Publisher 1886. Google Books. 6 Jan 2009. http://google.books/.

Mansfield, Donald H. *Flora of Steens Mountain*. Corvallis: Oregon State University Press, 2000.

Mansfield, George Rogers. "Geography of Southeastern Idaho." *Annals of the Association of American Geographers* 15:2 (Jun. 1925): 51–64. JSTOR. Utah State University Library, Logan: 16 Nov 2007. http://links.jstor.org/.

McKnight, Tom Lee. *Physical Geography: A Landscape Compendium.* Englewood Cliffs, NJ: Prentice-Hall, 1984.

McLaughlin, Steven P. "Natural Floristic Areas of the Western United States." *Journal of Biogeography* 16:3 (1989): 239–248. JSTOR. Utah State University Library, Logan: 14 Mar 2008. http://links.jstor.org/.

Nold, Robert. *Penstemons.* Portland, OR: Timber Press, 1999.

Neely, Elizabeth E., and Mary E. Barkworth. "Vegetation on Soils Derived from Dolomite and Quartzite in the Bear River Range, Utah: A Comparative Study." *Bulletin of the Torrey Botanical Club* 111:2 (Apr.-Jun. 1984): 179–192. JSTOR. Utah State University Library, Logan: 2 Sep 2009. http://links.jstor.org/.

Nevada Natural Heritage Program. (n.d.). Retrieved from http://heritage.nv.gov/images.htm.

Nicholls, Graham. *Alpine Plants of North America.* Portland, OR: Timber Press, 2002.

Oosting, Henry J. *The Study of Plant Communities: An Introduction to Plant Ecology.* San Francisco: W. H. Freeman, 1956.

Oregon Flora Project. (n.d.). Retrieved from http://oregonflora.org/checklist.php.

Ostler, W. Kent, and K. T. Harper. "Floral Ecology in Relation to Plant Species Diversity in the Wasatch Mountains of Utah and Idaho." *Ecology* 59:4 (Summer 1979): 848–861. JSTOR. Utah State University Library, Logan: 10 Dec 2010. http://links.jstor.org/.

Robertson, Leigh. *Southern Rocky Mountain Wildflowers.* Helena, MT: Falcon Publishing, 1999.

Rundel, Philip W., Arthur C. Gibson, and M. Rasoul Sharifi. "The Alpine Flora of the White Mountains, California." *Madroño* 55:3 (2008): 202–215. BioOne. 31 Oct 2011. http://www.bioone.org/doi/full/10.3120/0024-9637-55.3.202.

Scott, Richard W. "The Alpine Flora of Northwestern Wyoming." Thesis. University of Wyoming. 1966.

———. *The Alpine Flora of the Rocky Mountains: Vol. 1 The Middle Rockies.* Salt Lake City: University of Utah Press, 1995.

Shaw, Richard J. *Annotated Checklist of the Vascular Plants of Grand Teton National Park and Teton County, Wyoming.* Salt Lake City: Lorraine Publishing, 1992.

————. *Annotated Checklist of the Vascular Plants of Great Basin National Park and Adjacent Snake Range*. Logan: Utah State University, 1993.

————. *Utah Wildflowers: A Field Guide to Northern and Central Mountains and Valleys*. Logan: Utah State University Press, 1995.

————. *Vascular Plants of Northern Utah: An Identification Manual*. Logan: Utah State University Press,1989.

Snow, Neil. "Checklist of the Vascular Plants of the Southern Rocky Mountain Region (Version 3)." Colorado Native Plant Society. Jan. 2009. http://www.conps.org/plant_lists.html.

Taylor, Ronald J. *Sagebrush Country: A Wildflower Sanctuary*. Missoula, MT: Mountain Press Publishing Company, 1992.

Vale, Thomas R. "Pre-settlement Vegetation in the Sagebrush-Grass Area of the Intermountain West." *Journal of Range Management* 28:1 (Jan. 1975): 32–36. JSTOR. Utah State University Library, Logan: 14 Oct 2008. http://links.jstor.org/.

Vizgirdas, Ray S. *Useful Plants of Idaho*. Pocatello: Idaho State University Press, 2003.

Weber, William A. *Rocky Mountain Flora*. Boulder: University of Colorado Press, 1967.

Welsh, S. L., N. D. Atwood, S. Goodrich, and L. C. Higgins. *A Utah Flora*. Provo, UT: Brigham Young University Press, 1993.

Willard Basin, Box Elder County, Utah, 2011, Elevation: 9,280 ft.

Western Wildflowers - Quick Key

White Flowers	Flower Arrangement	Quick Identifier	Name	Page
	Along the stalk	Basal thin leaves	Elegant Death Camas (*Zigadenus elegans*)	18
		Found in wet, swampy areas	White Bog Orchid (*Platanthera dilatata*)	22
		Bottlebrush-like blossom	American Bistort (*Polygonum bistortoides*)	34
		Mat-forming	Mat Rockspirea (*Petrophytum caespitosum*)	44
		Rocky outcroppings	White Alumroot (*Heuchera bracteata*)	48
		Flowers on one side of the stalk	Side-flowered Miterwort (*Mitella stauropetala*)	52
		Flower petals shaped like a farmer's sickle	Sickletop Lousewort (*Pedicularis racemosa*)	58
	Cluster	Fern-like leaves	Common Yarrow (*Achillea millefolium*)	2
		Flower clusters resemble large cotton swabs	Pearly Everlasting (*Anaphalis margaritacea*)	4
		Gray-green feathery leaves	Douglas's Dustymaiden (*Chaenactis douglasii*)	6
		Several overlapping bracts at base of flower	Engelmann's Aster (*Eucephalus engelmannii*)	8
		Inflated flower base	Douglas's Catchfly (*Silene douglasii*)	10
		Shaded woodlands	Feathery False Solomon's Seal (*Maianthemum racemosum*)	16
		Red stems	Nuttall's Flaxflower (*Leptosiphon nuttallii*)	26
		Ladder-like leaves	Towering Jacob's Ladder (*Polemonium foliosissimum*)	30
		Whorled leaves	Whorled Buckwheat (*Eriogonum heracleoides*)	32
		Sticky stems with whorled leaves	Northern Bedstraw (*Galium boreale*)	46
		Colorful spots on flower petals	Spotted Saxifrage (*Saxifraga bronchialis*)	54
		Stream banks, with round scalloped leaves	Brook Saxifrage (*Saxifraga odontoloma*)	56
	Single	Palm-shaped leaves	Richardson's Geranium (*Geranium richardsonii*)	12
		Purple to brown band on each petal	Sego Lily (*Calochortus nuttallii*)	14
		Large papery flower petals	Tufted Evening-primrose (*Oenothera caespitosa*)	20
		Thistle-like, with large flowers	Flatbud Pricklypoppy (*Argemone munita*)	24

	Flower Arrangement	Quick Identifier	Name	Page
		Spiky leaves	Spiny Phlox (*Phlox hoodii*)	28
		Receding snowfields	Lanceleaf Springbeauty (*Claytonia lanceolata*)	36
		Succulent leaves, found only at high elevations	Alpine Springbeauty (*Claytonia megarhiza*)	38
		Marshy areas, with waxy round leaves	White Marsh Marigold (*Caltha leptosepala*)	40
		Toothed leaflets of three	Woodland Strawberry (*Fragaria vesca*)	42
		Deeply cleft flower petals	Woodland Star (*Lithophragma parviflorum*)	50
Yellow	**Flower Arrangement**	**Quick Identifier**	**Name**	**Page**
	Along the stalk	Blossoms resemble a feather duster	Prince's Plume (*Stanleya pinnata*)	100
		Clover-like leaves	Mountain Golden Pea (*Thermopsis rhombifolia*)	104
	Cluster	Thick feathery leaves	Sprawling Spring Parsley (*Cymopterus longipes*)	62
		Dandelion-like leaves and multiple flowers	Tapertip Hawksbeard (*Crepis acuminata*)	72
		Wiry looking	Showy Goldeneye (*Heliomeris multiflora*)	76
		Basal leaves and a hairy stem	Lambstongue Groundsel (*Senecio integerrimus*)	84
		Tall, with multiple flowering stems	Tall Groundsel (*Senecio serra*)	86
		Serrated arrow-shaped leaves	Arrowleaf Groundsel (*Senecio triangularis*)	88
		Shrubby	Canada Goldenrod (*Solidago canadensis*)	90
		Flowers found in the leaf axils	Puccoon (*Lithospermum ruderale*)	96
		Four petals on each flower	Western Wallflower (*Erysimum asperum*)	98
		Succulent	Spearleaf Stonecrop (*Sedum lanceolatum*)	102
		Umbrella-like flower clusters	Sulphur Flower Buckwheat (*Eriogonum umbellatum*)	112
		Feather-like leaves	Gordon's Ivesia (*Ivesia gordonii*)	118
		Leaves often cleft at the tip	Sulphur Paintbrush (*Castilleja sulphurea*)	122
		Petals look inflated	Tolmie's Owlclover (*Orthocarpus tolmiei*)	126
	Single	Dandelion-like leaves and single flower	Mountain Dandelion (*Agoseris glauca*)	64
		Woolly heart-shaped leaves	Heartleaf Arnica (*Arnica cordifolia*)	66

Quick Identifier	Name	Page
Large lobed leaves	Largeleaf Balsamroot (*Balsamorhiza macrophylla*)	68
Arrow-shaped fuzzy leaves	Arrowleaf Balsamroot (*Balsamorhiza sagittata*)	70
Rough leaves that clasp the stems	Rocky Mountain Dwarf Sunflower (*Helianthella uniflora*)	74
Huge flowers on a small plant	Tundra Hymenoxys (*Hymenoxys grandiflora*)	78
Smooth blue-green leaves	Owl's Claws (*Hymenoxys hoopesii*)	80
Tall, with drooping flower petals	Cutleaf Coneflower (*Rudbeckia laciniata*)	82
Spiky leaves	Stemless Mock Goldenweed (*Stenotus acaulis*)	92
Smooth shiny leaves	Mule's Ear (*Wyethia amplexicaulis*)	94
Backward-flaring flower petals	Yellow Avalanche Lily (*Erythronium grandiflorum*)	106
Drooping bell-shaped flower	Yellow Fritillary (*Fritillaria pudica*)	108
Leaves stick like velcro	Blazingstar (*Mentzelia laevicaulis*)	110
Moist areas, with succulent leaves	Hillside Buttercup (*Ranunculus jovis*)	114
High elevations	Alpine Avens (*Geum rossii*)	116
Sticky leaflets of three	Sticky Cinquefoil (*Potentilla glandulosa*)	120
Flower resembles a monkey's face	Seep Monkeyflower (*Mimulus guttatus*)	124
Goosefoot-shaped leaves	Goosefoot Yellow Violet (*Viola purpurea*)	128

Orange	Flower Arrangement	Quick Identifier	Name	Page
	Along the stalk	Bumps on the leaves	Gooseberryleaf Globemallow (*Sphaeralcea grossulariifolia*)	134
	Cluster	Flowers have blue anthers	Large-flowered Collomia (*Collomia grandiflora*)	136
		Very narrow leaves	Narrowleaf Paintbrush (*Castilleja linariifolia*)	138
	Single	Flowers normally have some red	Indian Blanketflower (*Gaillardia aristata*)	132

Red	Flower Arrangement	Quick Identifier	Name	Page
	Along the stalk	Fuchsia-like flowers	Garrett's Firechalice (*Epilobium canum*)	144
		Flared flower petals	Scarlet Gilia (*Ipomopsis aggregata*)	146
		Flowers on only one side of the stalk	Firecracker Beardtongue (*Penstemon eatonii*)	154

Color	Flower Arrangement	Quick Identifier	Name	Page
	Cluster	High-elevation succulent	King's Crown (*Rhodiola rosea*)	142
		Wavy forked leaves	Early Paintbrush (*Castilleja chromosa*)	150
		Multiple flowering stems with entire leaves	Scarlet Paintbrush (*Castilleja miniata*)	152
	Single	Shady locations	Western Columbine (*Aquilegia formosa*)	148

Color	Flower Arrangement	Quick Identifier	Name	Page
Pink	Along the stalk	Ladder-like leaves	Northern Sweetvetch (*Hedysarum boreale*)	168
		Square stems	Nettleleaf Giant Hyssop (*Agastache urticifolia*)	174
		Maple-like leaves	Mountain Wild Hollyhock (*Iliamna rivularis*)	180
		Palm-like leaves	Oregon Checkermallow (*Sidalcea oregana*)	182
		Rocky outcroppings	Wild Coralbells (*Heuchera rubescens*)	198
		Blue-green basal leaves	Palmer's Beardtongue (*Penstemon palmeri*)	202
	Cluster	Flower looks like a cotton swab	Rosy Pussytoes (*Antennaria rosea*)	158
		Tall, with three leaflets	Rocky Mountain Beeplant (*Cleome serrulata*)	164
		Hairy ladder-like leaves	Utah Milkvetch (*Astragalus utahensis*)	166
		Square stems	Wild Bergamot (*Monarda fistulosa*)	176
		Smells like an onion	Tapertip Onion (*Allium acuminatum*)	178
		Tall, with entire leaves	Fireweed (*Chamerion angustifolium*)	184
		Five flattened petals	Longleaf Phlox (*Phlox longifolia*)	186
		Pompom-like flower clusters	Cushion Buckwheat (*Eriogonum ovalifolium*)	188
		Looks like a dart	Darkthroat Shootingstar (*Dodecatheon pulchellum*)	190
		Moist areas	Parry's Primrose (*Primula parryi*)	192
		Drooping bulb-like flower	Old Man's Whiskers (*Geum triflorum*)	196
	Single	High elevations and cushion forming	Moss Campion (*Silene acaulis*)	162
		Low and looks like a steer's head	Steer's Head (*Dicentra uniflora*)	170
		Sticky stems	Sticky Geranium (*Geranium viscosissimum*)	172

		Paired flowers	Twinflower (*Linnaea borealis*)	160
		Hairy stalks	Cliff Anemone (*Anemone multifida*)	194
		Flower resembles a monkey's face	Lewis's Monkeyflower (*Mimulus lewisii*)	200

Purple	Flower Arrangement	Quick Identifier	Name	Page
	Along the stalk	Upright, with palm-like leaves	Silvery Lupine (*Lupinus argenteus*)	210
		Low growing, with palm-like leaves	Prairie Lupine (*Lupinus lepidus*)	212
		Bottlebrush-like flower	Purple Fringe (*Phacelia sericea*)	218
		Hooded flower	Western Monkshood (*Aconitum columbianum*)	226
		Resembles a small elephant's head	Elephanthead Lousewort (*Pedicularis groenlandica*)	230
		Flowers in whorls	Broadleaf Beardtongue (*Penstemon platyphyllus*)	232
		Flowers along one side of the stalk	Rocky Mountain Beardtongue (*Penstemon strictus*)	234
		Fuzzy flower throat	Whipple's Beardtongue (*Penstemon whippleanus*)	236
	Cluster	Flowers located below foliage	Ballhead Waterleaf (*Hydrophyllum capitatum*)	214
		Flowers located above foliage	Silverleaf Phacelia (*Phacelia hastata*)	216
		Clover-like flowers	Pale Monardella (*Monardella odoratissima*)	224
	Single	One to two rows of bracts directly below flower	Aspen Fleabane (*Erigeron speciosus*)	206
		Drooping bell-shaped flowers	Roundleaf Harebell (*Campanula rotundifolia*)	208
		Thick grass-like leaves	Western Iris (*Iris missouriensis*)	220
		Grass-like leaves	Douglas's Grasswidow (*Olsynium douglasii*)	222
		Drooping bell-shaped flowers with flared tips	Hairy Clematis (*Clematis hirsutissima*)	228
		Basal heart-shaped leaves	Hooked Violet (*Viola adunca*)	238

Blue	Flower Arrangement	Quick Identifier	Name	Page
	Along the stalk	Short	Lowland Larkspur (*Delphinium nuttallianum*)	260
		Tall	Western Larkspur (*Delphinium xoccidentale*)	262
		Tiered whorls of flowers	Wasatch Beardtongue (*Penstemon cyananthus*)	264

Flower Arrangement	Quick Identifier	Name	Page
Cluster	Sticky seeds	Meadow Forget-me-not (*Hackelia micrantha*)	244
	Tall, with bell-shaped flowers	Tall Mountain Bluebell (*Mertensia ciliata*)	246
	Short, with bell-shaped flowers	Leafy Bluebell (*Mertensia oblongifolia*)	248
	Airy	Prairie Flax (*Linum lewisii*)	252
	Ladder-like leaves	Showy Jacob's Ladder (*Polemonium pulcherrimum*)	254
	Tiny ladder-like leaves	Sky Pilot (*Polemonium viscosum*)	256
Single	Only at high elevations, with fuzzy leaves	Alpine Forget-me-not (*Eritrichium nanum*)	242
	Inverted bell-shaped flowers	Mountain Gentian (*Gentiana calycosa*)	250
	Shady areas	Colorado Columbine (*Aquilegia caerulea*)	258

Green

Flower Arrangement	Quick Identifier	Name	Page
Along the stalk	Tall	Green Gentian (*Frasera speciosa*)	268

Brown

Flower Arrangement	Quick Identifier	Name	Page
Along the stalk	Flowers look like little rabbits	Hare Figwort (*Scrophularia lanceolata*)	276
Single	Leaves are rough	Western Coneflower (*Rudbeckia occidentalis*)	272
	Drooping flowers	Leopard Lily (*Fritillaria atropurpurea*)	274

Index

The Authors

Richard Anderson is the Utah State University Botanical Center's nursery and greenhouse manager and oversees the plant introduction program. He resides in Malad, Idaho, with his wife and three children. He loves history and thoroughly enjoys spending time scouring the mountain terrain in search of the next botanical beauty.

JayDee Gunnell is a Horticulture Associate Professor for Utah State University Extension in Salt Lake County, Utah. He also oversees the arboretum development at the Utah State University Botanical Center. He resides in Layton, Utah, with his wife and four children. He enjoys any activity that finds him out of doors.

Jerry Goodspeed is a Horticulture Associate Professor for Utah State University Extension in Weber County, Utah, and co-directs two university-funded botanical gardens. He and his wife reside in Kaysville and enjoy spending time with their children and grandchildren. He is also known for his dry wit and humor that he finds in just about any situation.

Left to right: JayDee Gunnell, Richard Anderson, Jerry Goodspeed